From Sunlight to Electricity

From Sunlight to Electricity

A practical handbook on solar photovoltaic applications

(SECOND EDITION)

Foreword
R K Pachauri

The Energy and Resources Institute

ISBN 978-81-7993-156-1

This document was prepared by Mr Shirish Sinha, Mr Anand Shukla, Ms Preeti Malhotra, Mr Ibrahim H Rehman, and Ms Nandita Hazarika of the Rural Energy Group at TERI, New Delhi.

The book has been revised and updated by Dr Suneel Deambi. Dr Deambi is a Consultant (Renewable Energy) and a prolific writer on energy–environment issues.

Published by	Tel. 2468 2100, 4150 4900
The Energy and Resources Institute (TERI)	Fax 2468 2144, 2468 2145
TERI Press	India +91 • Delhi (0)11
Darbari Seth Block	E-mail teripress@teri.res.in
IHC Complex, Lodhi Road	Website www.teriin.org
New Delhi 110 003 / India	

Printed in India

Contents

Foreword

The PV (photovoltaic) technology programme in India has undergone several changes in terms of technology, approaches to promotion, financial arrangements, and so on. Numerous training programmes have been conducted in the past to equip extension agents, workers of NGOs (non-governmental organizations), and other stakeholders with advancements made in the PV sector. As part of such efforts, numerous documents have also come out as informational packages, dealing with specific issues in the sector — technical, financial, operational and maintenance, and so on. However, there is a lack of a comprehensive document that provides all information related to applications of PV technology, approaches to dissemination, financial arrangements, and institutional and social factors in technology acceptance, beneficial to implementing agencies, financial organizations, and other stakeholders in the PV sector in India.

The present book is an attempt in this direction. It compiles information that gives the reader an overall understanding of the PV sector in India, designs and applications of specific devices and related benefits, finance, policies, and programmes. The book is not only a compilation of information, but also gives a step-by-step methodology for designing PV systems for appropriate applications in rural areas. The book has been written in simple language with numerous illustrations such that readers not very familiar with the PV sector find the contents easy to understand. For others, the book aims at providing guidelines on sizing and installing PV systems, which they will find very useful in their work. This volume is expected to be of interest and importance to all the stakeholders in the PV sector, namely, policy-makers, government officials, NGOs, academic and research organizations.

R K Pachauri
Director-General, TERI, and
Chairman, IPCC (Intergovernmental Panel on Climate Change)

Preface

Electricity is one of the most vital energy sources for economic activities. The economic development of a country by and large depends on its efficient supply. It can help in transforming the way people live and work. In India, the rural electrification programme started in the 1950s with the aim of promoting economic development and improving the quality of life in rural areas. However, the impact of the programme has been very low. Scattered households in rural areas and the high cost of extending grid electricity supply to these areas have been identified as reasons. Installations of SPV (solar photovoltaic) systems for small-scale electricity generation directly from sunlight can help improve activities in the domestic, health care, agriculture, education, and micro-enterprises sectors. SPV systems have been considered as one of the best options for rural infrastructure projects in electrified/unelectrified areas.

This document compiles information on several aspects, ranging from the history of the technology to its principle, components, applications, advantages, and its limitations. It gives a step-by-step methodology of designing a PV system for appropriate applications in rural areas. The specific applications considered are domestic lighting systems and pumping systems. It also carries lists of system manufacturers and integrators, financial institutions and intermediaries, and state nodal agencies with their communication addresses.

The document contains seven chapters. Chapter I introduces PV technology, its principles, benefits and limitations, and the efforts of the government to popularize the technology in rural areas. Chapter II explains the components of a PV system. Chapter III discusses the applications of various PV systems, namely, domestic lighting systems, solar lanterns, street lights, PV pumps, and refrigeration. Chapter IV deals with designing of a PV system for various applications. In Chapter V, the costing of PV systems has been described. Other information includes market prices of systems/ components, central government subsidy, credit facility extended by IREDA (Indian Renewable Energy Development Agency), and the procedure to process soft loans. Chapter VI gives valuable tips on the operation and maintenance of PV systems. Chapter VII provides an overview of the SPV programme in India, and chapter VIII provides overview of the international PV programme and market.

The document helps a novice understand the technology and appreciate its various applications. It will also help those who are already familiar with the technology, as it carries guidelines on sizing and installing PV systems in rural areas for appropriate applications.

Overview of solar photovoltaic programme worldwide

Background

Photovoltaics, or PV, is a solar power technology that uses solar cells to convert incident sunlight directly into electricity with zero emissions. By now, PV technology has established itself as one of the best solutions to bring forth flexible and long-term solutions for rural electrification in the poorest areas of the world. Even though electrification began more than a century ago, there are still around 1.6 billion people without access to electricity in developing countries. PV electricity could contribute in reducing drastically this whopping figure and thus improve the quality of life in these regions.

The first few years of the 21st century have witnessed considerable development in the area of PV generation-cum-utilization. There are various reasons for this, including the much talked about global warming, consequent upon which large-scale deployment of solar panels has taken place in countries like Germany and Japan. The growth rate of 30%–40% during the last

Solar panel

few years is indeed a great incentive for investment the world over. The global market for solar PV has grown substantially in 2007 and early 2008. Entrepreneurs, venture capitalists, and big industrial houses have shown their eagerness to set up industries in the PV area. Developing countries like India and China are no exception to that, and attaining economies of scale in the production setup may not be a distant reality anymore.

Key issues

The key issue that has an impact on the large-scale use of PV technology is its affordability. It still suffers from the tag of having a high initial capital cost. To remedy this situation, intensive R&D (research and development) efforts are under way across the world to enrich the existing materials, besides finding new materials, processes and device structures. This is being done with the ultimate objective of reducing the use of expensive silicon material and raising the solar-to-electric conversion efficiencies of the solar cell—a basic unit of the PV system. The resultant effect could be in the form of a reduced per peak watt price of the solar system. Thin film amorphous silicon did emerge as a viable alternative many years ago, but despite initial market success, the share of thin film PV technology fell drastically to just 8%. However, the market share of thin film technologies is expected to reach up to 20%, mainly due to some innovative developments with regard to materials like cadmium telluride. In India, too, a few major thin film manufacturing facilities are under active consideration, which may finally ease the pressure on the constrained supply of silicon wafers in the international market.

The grim reality is that the growth in the global demand for solar panels is still fully dependent on and limited by local subsidy programmes. The market segments that are not dependent on subsidies such as autonomous PV applications represent an extremely small portion of the world market. It is expected that the new generation solar modules could bring about new markets that are independent of the government subsidies.

Developed markets

As regards the existing PV markets in 2007, Germany edged past the rest in terms of a sizeable market share of just under 50% due to an incentive model better known as the feed-in-tariff system. In contrast, the Japanese market share grew in 2006, but fell again, representing just about 8% of the global market in 2007. It was mainly due to the discontinuation of a national subsidy scheme in 2005. The PV market in the USA was dominated by subsidy programmes, especially in New Jersey and California. The PPA (power purchase agreement) model became quite popular in the USA, being especially dependent on tax credits. The PV market in Spain showed spectacular growth in 2007, again by virtue of the feed-in-tariff scheme. Other countries with excellent growth prospects are Italy, South Korea, France, Greece, and also relatively smaller markets like Belgium, Czech Republic, the Netherlands, and Portugal.

Developing markets

Around 150 countries are typically categorized as developing countries due to the lack of a high degree of industrialization, infrastructure, and capital investment. Of late, there has been considerable growth in the PV market of many of these countries, driven essentially by China and India. China is pursuing an aggressive policy of creating large-scale infrastructure along the complete PV chain and may soon start delivering cheap solar modules and accompanying product range.

Solar PV fits the needs of developing countries, as it turns out to be cheap for rural areas, thus avoiding the infrastructure investment. Labour costs for system installation are also low. The market share of PV systems in the developing countries is expected to go up from 800 MW_p (megawatt peak) in 2007 to 1900 MW_p by 2010, with China turning out to be the largest PV market. The solar PV market is expected to take an upward swing in other regions of the developing world. Amongst the developing countries, India took a fairly early lead in the large-scale development of a solar PV programme. Incidentally, its demonstration programme was regarded as one of the largest PV programmes on the global level.

Value-based applications

Solar systems have an added value in rural areas, especially in developing countries. Lighting systems, together with systems used for water pumping, vaccine refrigeration, and battery charging, continue to be in good demand in countries like India, Nepal, Sri Lanka, Bangladesh, Indonesia, and Senegal. In addition, the use of solar electricity for remote applications is quite common in the telecommunications field, especially to link remote areas to the rest of the country. In fact, repeater stations for mobile phones powered by PV or a hybrid system have a large potential.

In totality, rural electrification via PV brings well-being and creates the conditions for economic development where it is most needed. The scenario is still very grim for off-grid rural communities in developing countries, which still depend on kerosene or paraffin lamps for lighting. Kerosene oil is heavily subsidized but has limited availability in some rural areas. Worldwide, around 1.5 million people die annually from indoor air pollution due to lighting and cooking. A positive fallout is that it has boosted the sales of products like solar lanterns and solar home systems, or SHS. These have been largely sold in developing countries, including India, both under the subsidy programmes and via the micro-financing route. The micro-finance creditors collect payments as often as on a weekly basis from those who normally struggle to set some money aside.

Path-breaking initiatives

Intensive deliberations on the ways and means to reduce the cost of solar PV systems have been conducted to give a fillip to PV programmes in developing countries. It also includes using more energy-efficient lamps, along with increasing the life of the battery, besides innovative financing mechanisms.

As of now, PV systems like grid-interactive power plants and BIPV (building integrated PV) are making gradual inroads in developing country markets as well. In India, the MNRE (Ministry of New and Renewable Energy), for example, is offering sops for promoting a wide range of systems like solar road studs, solar signages, traffic

An energy-efficient building with building integrated PV

lighting systems, and street lighting systems, mainly through state municipal bodies.

The World Bank has recently announced a competition for the private sector to innovate the best value low-carbon light source for households in Africa to stimulate an estimated African market of around $17 billion for off-grid lighting. Solar products based on the LED (light emitting diode) can bring about remarkable developments in the developing world in consonance with an expected price reduction in the solar modules. The Lighting a Billion Lives, or LaBL, initiative kickstarted by TERI (The Energy and Resources Institute) in 2008 aims to empower people still reeling under darkness—India is home to around 25% of such people.

Conclusion

The way forward is to make hay while the sun shines through the use of low-cost technology, well-designed and need-based systems, easy financing, and importantly, a truly sustainable O&M (operation and maintenance) mechanism. Realizing such a widespread objective calls for a hand-in-glove approach between the developed and developing countries. After all, our planet should become a good living place for everyone by utilizing the freely available and abundant sunshine.

Components of photovoltaic systems

Introduction

An SPV (solar photovoltaic) system is an independent energy-generating system, which provides power for different types of devices. PV systems are classified into two types—one provides power generated directly from solar modules, and the other generates and stores power, which is utilized as and when required. Thus, the usage of power generated either directly or in a stored form depends on the type of application. Figure 1 illustrates the two types of system configurations. In the first configuration, power generated is used to operate the load directly. In the second, power generated is stored and used according to the demand, that is, during low sunshine and/or non-sunshine hours. The most commonly used storage systems are batteries. Charge controllers are added to the system to prevent overcharging or overdischarging (also called deep discharging) of the batteries, which would

Solar photovoltaic panels

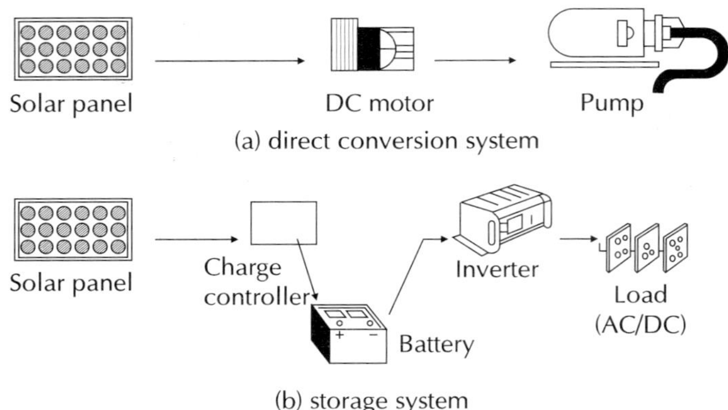

Figure 1 PV system configurations

otherwise affect their performance and life. An inverter is used for running of any AC loads, for example, an AC submersible pump.

Irrespective of the system configuration, a PV module is the basic component of any system. The main building blocks of a PV module are solar cells. All components in the system, other than the module are termed as BoS (balance of system). The BoS consists of batteries, charge controllers, inverters, and other accessories like wires, cables and protective fuses, and support structure. The final configuration of a PV system normally depends upon the site location and specified load requirements.

Silicon wafer

Silicon is the basic raw feedstock material for the PV industry. The primary source of silicon is quartzite, found in ordinary sand. It takes a lot of energy to make a pure silicon wafer, on which a solar cell is grown. Two most common techniques to produce wafers are the chzrolaski (CZ) and floating zone (FZ) methods. Silicon thus processed does not come cheap, so it is important to use lesser quantity of silicon. The thickness of the wafer is quite important in determining the cost of a solar cell.

Silicon is purified, melted, and crystallized into ingots. The ingots are then sliced into wafers to make individual cells. The average thickness of wafer has been reduced from 0.32 mm in 2003 to about 0.18 mm by 2007.

Some quantity of precious silicon material is also lost during wafer production. A way out is to use ribbon sheet technology, which involves pulling thin wafers from the melt, thus eliminating need for sawing of wafers. The Indian PV industry now makes use of silicon wafers in the thickness range of 210–240 microns as against 350–400 microns till a few years back.

Solar cells

The fundamental power unit of a PV system is a solar cell. A solar cell is grown on a thin wafer of silicon, which is chemically treated and processed at high temperatures. A p–n (positive–negative) junction is fabricated, across which voltage develops when sunlight falls on these cells (Figure 2). The four most common types of silicon-based PV cells are the following.

1   Single crystal silicon
2   Polycrystalline silicon
3   Ribbon silicon
4   Amorphous silicon

Individual cells are attached to a base called a back-plane, which is usually a layer of metal used to physically reinforce the cells and provide electrical contact at the bottom. Polycrystalline cells are manufactured by a similar process using relatively less pure

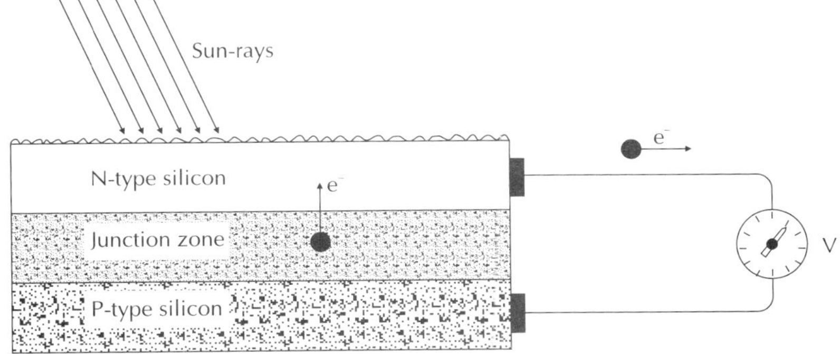

Figure 2 Working principle of a solar cell

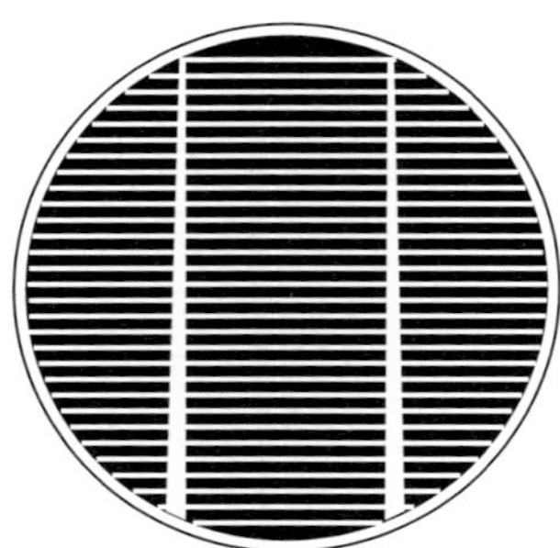

Figure 3 A crystalline silicon solar cell

silicon. In case of ribbon type PV cells, molten silicon is drawn into the form of a ribbon. Most PV cells (more than 90%) are of the crystalline silicon type (Figure 3).

Thin film cells

As mentioned above, pure silicon is a very expensive material. So, continuous efforts are on to produce cells with very little or no quantity of expensive silicon. This type of cell technology is usually known as thin film technology. Thin film modules are made by depositing very thin layers of photosensitive materials on to a low-cost backing. Typical inexpensive substrates used for the purpose are made of glass, stainless steel or even plastic. Three types of thin film modules are commercially available at present.
1   Amorphous silicon
2   Cadmium telluride
3   Copper indium diselenide

These have active layers in the thickness range of less than a few microns and are suitable for large-scale production. One of the unique advantages of amorphous silicon is that it can be moulded into many different shapes. However, thin film cell technologies possess lower solar-to-electric conversion efficiencies in comparison to crystalline silicon cell technologies. Table 1 shows the available cell and module efficiencies.

Crystalline silicon is the most dominant cell technology at present. It has a market share in excess of 90%. Thin film technologies, on the other hand, account for less than 10% of market. Table 2 shows the respective shares of available cell technologies in 2006.

The PV market is always looking for some dramatic breakthroughs in any of the available cell technologies. It is quite clear from Tables 1 and 2 that crystalline silicon cells always edge out the thin films due to several advantages they have over amorphous silicon. Table 3 shows the relative advantages and disadvantages of thick (crystalline silicon) and thin film (amorphous silicon for example) technologies.

Table 1 Currently available cell and module efficiencies

Technology	Cell efficiency (%)	Module	Remarks
Single crystal silicon	16–17	13–15	Highest efficiency but more expensive
Polycrystalline silicon	14–15	12–14	Slightly less efficient but no distinct cost advantage
Amorphous silicon		6–7	Very low efficiency but cheaper fabrication cost
Cadmium telluride		8–10	Moderately efficient, just beginning to make market impact
Copper indium diselenide		10–11	Most efficient amongst rest of thin films, negligible market share

Source Greenpeace 2007

Table 2 Shares of available cell technologies in 2006

Cell technology	Share (%)
Single crystal silicon	43.4
Polycrystalline silicon	46.5
Ribbon sheet (crystalline silicon)	2.6
Amorphous silicon	4.7
Cadmium telluride	2.7
Copper indium diselenide	0.2

Source Greenpeace 2007

Table 3 Comparison between thick (crystalline silicon) and thin film (amorphous silicon for example) technologies

Crystalline silicon	Amorphous silicon
High quality cells with no change in performance levels	Not very stable under outdoor light conditions (loses some power initially)
High solar-to-electric conversion efficiency (14%–17%)	Very low efficiency
Long life (about 25 years)	Short life (<10 years)
Occupies market share of more than 90%	Market share less than 5%

Solar photovoltaic module

The power output of a single solar cell is too low to operate most electrical devices. Most single junction cells produce a voltage of

A solar photovoltaic module

about 0.5–0.6 V, regardless of the surface area of the cell. This voltage remains fairly constant with changing light intensity. However, the current in a cell is almost directly proportional to light intensity and size. Therefore, many cells are connected in series to increase the voltage. Several series of cells can also be interconnected in parallels to increase the power output.

Solar cells are extremely fragile. To protect them from damage, they are hermetically sealed between a top layer of glass or clear plastic, and a bottom layer of plastic or a combination of plastic and metal. An outer frame is attached to increase the strength, and this whole package is called a PV module. At the back of the module, a junction box is provided to extract electricity (Figure 4). Depending upon the load and battery requirements, the modules are connected in series/parallel combinations and mounted on a metallic frame. Such aggregates of PV modules form a PV array. In India, the majority of the PV modules available in the market are either monocrystalline or polycrystalline silicon type.

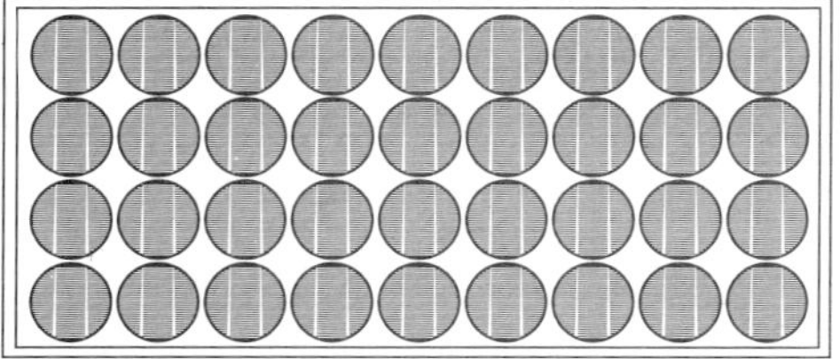

Figure 4 An SPV panel

In general, crystalline silicon modules are robust, reliable, and weatherproof. Table 4 shows the average area requirements for different types of solar modules.

Table 4 Average area requirements for different types of solar modules

Type of module	Area requirement (per kW_p)
Single crystal silicon	~ 7 m²
Polycrystalline silicon	~ 8 m²
Amorphous silicon	~ 15 m²
Cadmium telluride	~ 11 m²
Copper indium diselenide	~ 10 m²

kW_p – kilowatt peak

It is clear from the above that higher the conversion efficiency of a cell material, lower is the area requirement. The thin film panels based on amorphous silicon material in particular need more area for installation purposes. It may not be an issue in remote rural areas, but in urban areas, roof or land area is generally expensive. A way out is to integrate the thin film panels within a building envelope using its advantage of getting moulded in many different shapes unlike the crystalline silicon panels.

Power output from a solar module

The SPV modules are rated in terms of W_p (peak watt) units. Modules supply power depending on the amount of sunlight received by them. Since the amount of sunlight received varies, the power, or watts supplied, also varies. To accommodate the variation in the power production, watt peak is used. It is the maximum rated output of a solar cell, module or array under STC (standard test conditions), usually 1000 watts/m² (0.645 watts per square inch) of sunlight with other conditions such as temperature specified. Typical rating conditions are 25 °C and 1 m/sec of wind speed.

The amount of solar radiation (energy) received on a given surface area in a given time is also known as solar insolation or even sun-hours per day for simplicity. It is commonly expressed as watts per square metre (W/m²) or kilowatt-hours per square metre per day (kWh/m²/day) or kilowatt hours per year per kilowatt peak rating. The solar insolation of a region indicates the total hours for which the rated 'peak watt' from a module can be obtained. In India, the most commonly used modules are rated as 37 W_p

(generally used in domestic lighting and pumping systems), and 10 W_p for portable lighting systems (solar lanterns). However, modules with power capacity of 2–250 W_p are now being produced indigenously by several manufacturers. The average solar insolation in India is 5.5 kWh/m^2/day.

The power output of a crystalline silicon module decreases with increase in ambient temperature. Against each unit (W_p) of rated power, the module should produce about 0.85 Wh of electrical energy for each unit (kWh/m^2/day) of solar insolation. For example, a 37-W_p module would produce 37×0.85 Wh = ~32 W of electricity at STC in India.

To sum it up, the following factors reduce the power output from a solar module.

- Peak watt rating of the module—larger modules yield more power than smaller modules.
- Intensity of solar radiation—more power is produced under bright sunlight than when it is cloudy. For example, the current produced in the latter case may just be 20%–30% of that generated in bright sunlight.
- Time of the year—it also brings down the solar output. For example, solar intensity is reduced in winter.
- Hours of available sunlight—time of the year has a definite effect on the duration of light available to a module

At present, it is a common practice for reputed crystalline silicon module manufacturers to guarantee a power output of 80% of the nominal power even after a period of 20–25 years.

Balance of system

All system components other than the PV module are termed as the BoS. The most critical component of the BoS is the battery. It is used when the system is to be operated during non-sunshine hours or if the loads operate on DC. The other components generally include charge controller, inverter, support structure, and wiring and cabling.

Batteries

The battery is the most common type of storage device in SPV systems. Batteries store the electrical energy generated by the modules during the day, and deliver it during cloudy days or

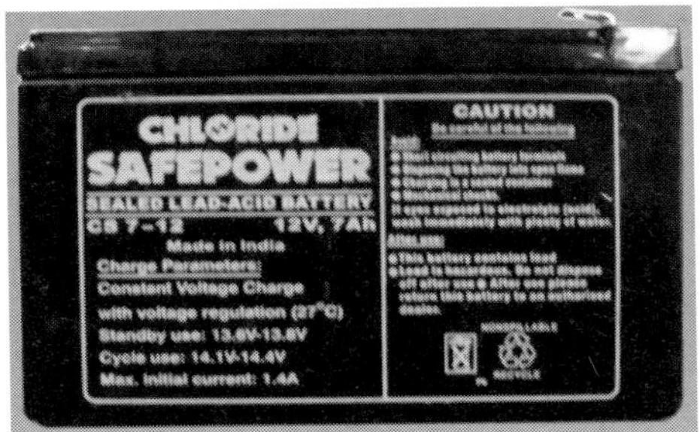

Battery to store electricity

non-sunshine hours. The battery needs DC for charging. The electricity generated from solar cells is also DC, which can be easily stored in the battery, without involving any intermediary process (Figure 5). Battery capacity is generally expressed in terms of V (voltage) and Ah (ampere-hour), for example, a 12-V battery at 40 Ah. For solar applications, a battery needs to be capable of being discharged hundreds or even thousands of times. That is why a solar battery is commonly known as a deep-cycle battery.

The batteries are recharged to 100% capacity during the charging phase of each cycle. A cycle is described as an interval that includes one period of charging and discharging. However, the batteries must not be completely discharged (that is, below 80%) during each cycle.

If a charge controller is not included in the system, oversized loads or excessive use can drain the batteries' charge to the point where they can be damaged and have to be replaced. Similarly, the controller regulates overcharging of the batteries, which can be damaged during times of low or no use or long periods of full

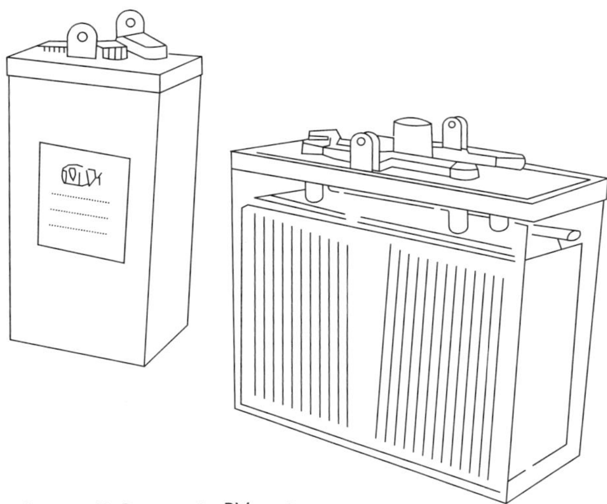

Figure 5 Battery in PV systems

sunshine hours. Two types of batteries are generally used with the PV systems, that is, lead–acid batteries and nickel–cadmium batteries. In India, however, the most commonly used batteries are lead–acid batteries.

Lead–acid batteries are further classified into flooded and dry cell types. In most of the domestic or large-scale applications of PV systems, flooded lead–acid batteries are used. Dry cell batteries are used in portable lighting systems such as solar lanterns. Table 5 lists the advantages and disadvantages of sealed batteries, usually used in low-power devices like a solar lantern.

Table 5 Advantages and disadvantages of sealed batteries

Advantages	Disadvantages
Need no maintenance	More expensive than other battery types
Are spillproof	Need more accurate charging control
Most appropriate where solar systems need to operate for long periods without any maintenance	Have a short life usually at high temperatures

The dry cell batteries are normally of 1–10 Ah, while the flooded batteries are available from 40–1000 Ah or even more. The following are critical performance parameters for PV batteries.

- Storage capacity
- Charging and discharging efficiency
- Self-discharge
- Capability to operate in different state of charging modes
- Operation and maintenance procedures

The battery efficiency for most PV applications is 85%. Depending upon the load size, one or more batteries may be required. Such an aggregate of batteries is called a battery bank. It may be designed at 24, 48 or 96 volts in case of small power plants. The lifespan of a battery mainly depends on the battery management and the user's attitude.

Charge controller

Charge controller, as the name suggests, is a device used to control the amount of charge flowing in and out of a battery. The battery is connected to the PV array through a charge controller. A charge controller is a solid-state electronic component, which controls the electricity generated by a solar module (Figure 6). It acts like a voltage regulator and is placed between the PV module and the battery. The primary function of a charge controller in a stand-alone PV system is to manage the rate and amount of charging of batteries. Thus, it protects the battery from overcharging or deep discharging. Absence of charge controllers may result in shortened battery life and decreased load availability.

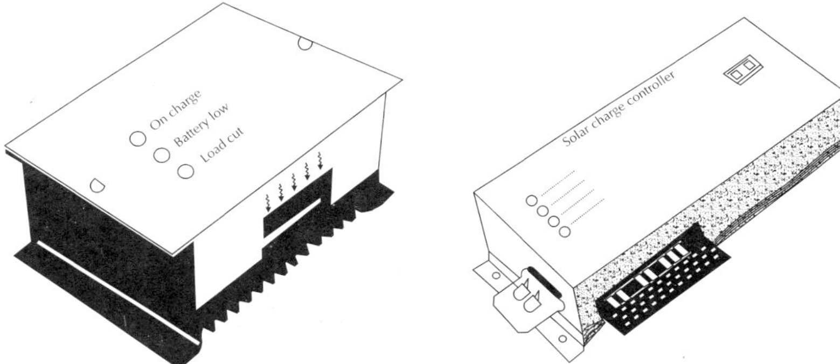

Figure 6. Charge controller

Solar charge controllers are specified by the system voltage they are designed to operate on and the maximum electrical charge they can safely handle. The system voltage is normally 12 V or 24 V or occasionally 48 V. The maximum current is determined by the number and size of modules. A single module would need a charge controller of between 4 A and 6 A, while larger arrays may need charge controllers of 40 A or even more.

Inverters

Photovoltaic systems work more efficiently with DC loads (as there are no conversion efficiency losses). However, certain applications or AC loads can also be operated with solar electricity using an inverter. Inverters are electronic components, which convert DC power generated by a solar array into AC compatible with the local distribution network. DC has a current flow in one direction only, while AC rapidly switches the direction of current flow back and forth. In a typical PV system, inverters are placed after the batteries and before the loads. These should be placed reasonably close to the battery bank but not inside the battery enclosure to avoid problems such as overheating. There are many different types of inverters available for use with solar power applications.

Inverters are produced in different ranges of power categories ranging from a few hundred watts through the widely used range of few kW_p (3–6 kW_p). There are also inverters designed for large-scale solar power systems of 100 kW_p and more. The two most common types of inverters are known as modified sine wave and sine wave types. Modified sine-wave-based inverters are cheaper in comparison to pure sine wave. The latter type is generally used for grid-interfaced applications. In totality, the main function of an inverters is to transform the low-voltage DC generated by the solar module into mains voltage AC.

Blocking diode

The blocking diode is yet another type of charge controller device. This acts as a check valve to prevent discharge of batteries (reverse flow of current to the solar cells) through PV module during night or during insufficient sunlight.

Other components

Other components of PV systems include electrical connecting and physical mounting equipment. Electrical connecting equipment generally comprises electrical wires and cables connecting PV modules, and from modules to batteries, inverters, and loads. The physical mounting equipment comprises metallic frames, nuts, bolts, and clamps, which are used either for rooftop mounting of modules or ground-based installation.

Reference

Greenpeace. 2007
Solar Generation IV-2007
Details available at http://www.epia.org/fileadmin/EPIA_docs/publications/epia/EPIA_SG_IV_final.pdf, last accessed on 5 September 2008

Applications of photovoltaic systems

Introduction

PV (photovoltaic) systems have many different applications in the off-grid areas. Present day systems are much more reliable, mainly due to vast improvements in the technology. Thus, the remotely located end user has benefited in the process. In the developing world, there is a huge demand for solar lighting systems, followed closely by demand for water pumping systems—both for drinking and irrigation purposes. PV systems are also useful for storing vaccines, besides lighting up health clinics in rural areas. In general, PV technology plays an important role in linking far-flung villages with national development initiatives. That is not all; there are many more applications of PV across the domestic, commercial, and industrial sectors. Figure 1 summarizes the applications of PV systems.

Lighting systems

Lighting is by far one of the most common applications of PV systems in remote rural areas. Lighting systems are generally of two types—indoor and outdoor. Indoor lighting systems are further classified into fixed and portable. The fixed indoor lighting system is commonly known as SHS (solar home system). Portable indoor lighting systems are generally the solar lanterns and solar torches. Outdoor lighting systems include street lights, garden lights, and pathway lights. However, with an increasing demand for electricity for entertainment needs (such as television), the definition of SHS has changed from being a two-light point system with an additional provision for operating a portable black and white TV, radio or fan.

A solar lantern in use

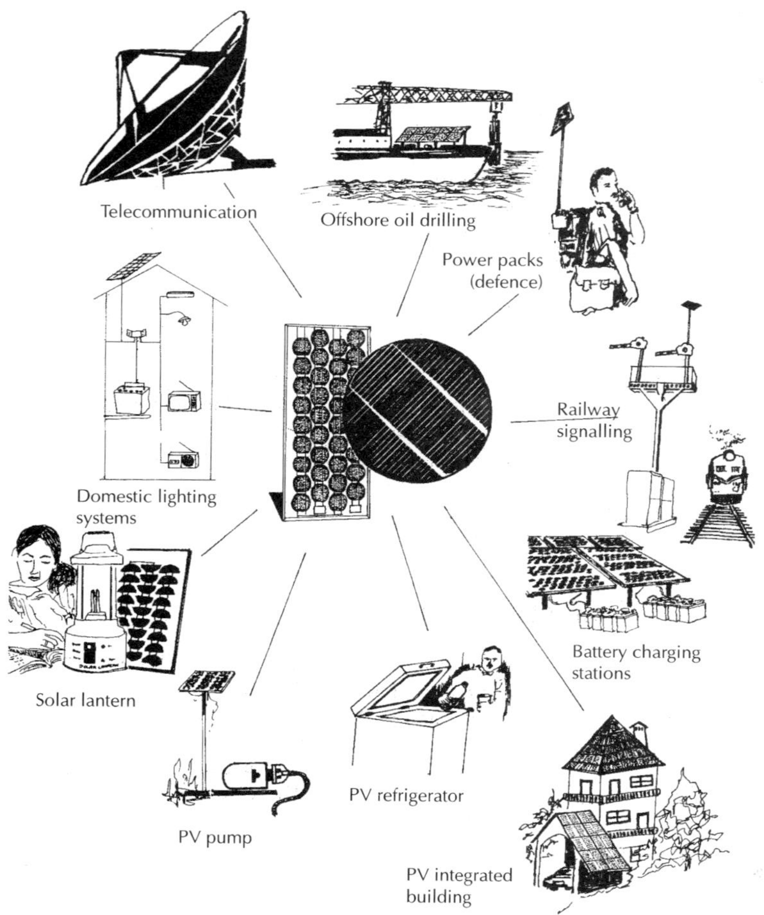

Figure 1 Applications of PV systems

Solar home systems

SHS, as the name indicates, is designed to meet the domestic
lighting requirements. The most widely used SHS is a 37 W_p system
with provision for two light points. All the systems installed under
the MNRE (Ministry of New and Renewable Energy) programme
have to adhere to the following five configurations for SHS
(Table 1). There is an incentive scheme in the form of direct cost
subsidy for the buyers of such a system.

Table 1 MNRE specifications for SHS configurations

Configuration	Module voltage (W$_p$)	Battery	Load	Daily usage (hours)
I	18	12 V, 20 Ah	1x CFL 9/11 W	3–4
II	37	12 V, 40 Ah	2x CFL 9/11 W	3–4
III	37	12 V, 40 Ah	1x CFL 9/11 W 1x DC fan(<20 W)	1 light, 1–2 hrs 1 fan, 1–2 hrs
IV	2x37 or 1x74	12 V, 75 Ah	2x CFL 9/11 W 1x DC fan(<20 W)	2 lights, 2–3 hrs 1 fan, 2–3 hrs
V	2x37 or 1x74	12 V, 75 Ah	4x CFL 9/11 W	3–4 hrs

W$_p$ – watt peak; Ah – ampere-hour; V – volt; CFL – compact fluorescent lamp, hrs – hours

Source www.mnes.nic.in

Various other designs of SHS can be made available by the system manufacturers as per the actual load requirements. A user may purchase such systems on an outright cost basis or through soft loan scheme of IREDA (Indian Renewable Energy Development Agency). The key components of an SHS (Figure 2) are briefly described below.

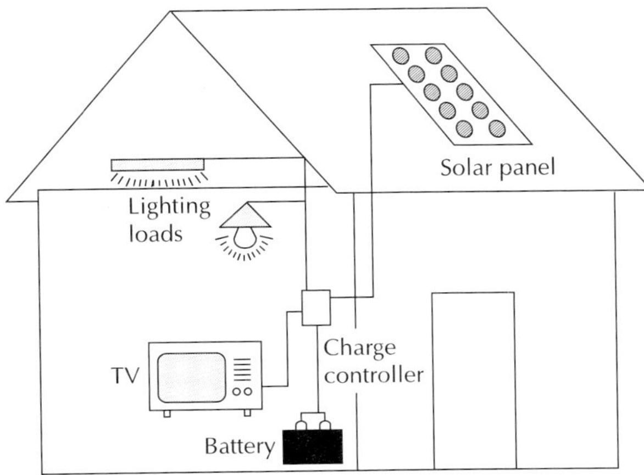

Figure 2 Key components of a solar home system

PV module

The PV module generally consists of crystalline silicon solar cells. Its peak power capacity varies from 18 W$_p$ to 74 W$_p$, and this can be increased depending upon the load. The module should be installed at an optimal angle (generally the latitude of a place) and direction, either on the rooftop, or at a place where it gets the maximum sunshine.

Battery

The batteries used in SHS are flooded electrolyte type lead–acid positive tubular plate, low-maintenance lead–acid batteries. It will have a minimum rating of 12 V, and 20, 40 or 75 Ah (at C/10) discharge rate, depending on the configuration chosen. Seventy-five per cent of the rated capacity should be between fully charged and load cut-off conditions.

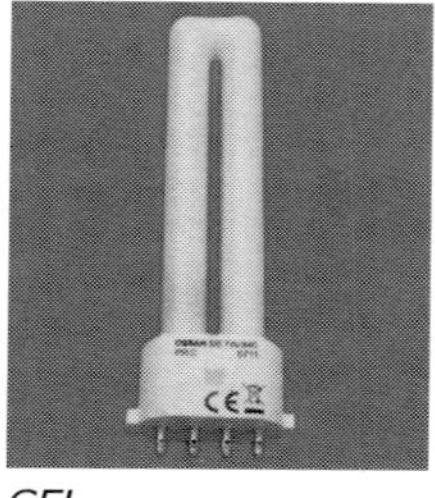

CFL

Lamps

CFLs (compact fluorescent lamps), either 4-pin or 2-pin types, are used. The ratings of these CFLs are either 9 W or 11 W. In case of 4-pin type CFLs, a preheating circuit is provided. The CFL is placed in an assembly, often called luminaire, with a reflector on its back. Lamps are installed in a base-up position. The light output from the lamps should be around 600±5% lumens for a 9-W CFL and 900±5% for an 11-W CFL.

Other loads

The other loads used with an SHS could be a DC fan, portable black and white TV set, and so on. The loads of both these are in the range of 18–20 W.

Electronics

The inverter should be of a quasi-sine wave/sine wave type. Its frequency must lie between 20 kHz and 35 kHz. The total electronic efficiency should be at least 80%. There should be no blackening or reduction in the lumen output by more than 10% after 1000 ON/OFF cycles. (Two minutes ON, followed by four minutes OFF is one cycle.) The idle current consumption should not be more than 10 mA (milliampere). Electronics should operate at 12 V and should have temperature compensation for proper charging of battery round the year. A blocking diode is also provided to stop the reverse flow of current to PV module from batteries. The short circuit condition is prevented by using a fuse. Necessary lengths of wires/cables, switches suitable for DC use and fuses should be provided.

Electronic protections

There must be full protection under no-load conditions—when the lamps are removed and the system is switched ON. The battery runs the risk of getting overcharged or deeply discharged. So, a system needs to be protected against these two conditions as well. In addition, there should be complete protection against accidental short circuit and reverse polarity.

Mechanical components

A solar module is the most delicate part of a PV system. It must be mounted properly on a metallic frame structure having corrosion-resistance paint. Such a frame must have a provision to adjust its angle of inclination to the horizontal between 0° and 45°. The purpose is to install the module at a specified tilt angle. The battery emits fumes during the charging process, so it should be kept in a vented metallic/plastic/wooden box.

Other features

Solar systems are generally installed in remote rural areas. So, it should have easy-to-follow instructions. A green LED (light emitting diode) shows the charging while a red LED indicates a deeply discharged condition of a battery.

Solar lantern

A solar lantern is a single light point portable lighting system. Apart from a small-capacity PV module, it consists of a lamp, battery, and electronic components, all placed in a casing, either made of metal, plastic, or even fibreglass. Because of its portability, it is quite useful both in indoor and outdoor environments. Most solar lanterns radiate 360 degrees of light output (Figure 3). The limited unidirectional solar lanterns are no longer in use under the government-supported programmes. There are a few configurations of a solar lantern approved by MNRE. Table 2 shows a few such configurations with different capacities of solar modules and lamp types.

A solar lantern

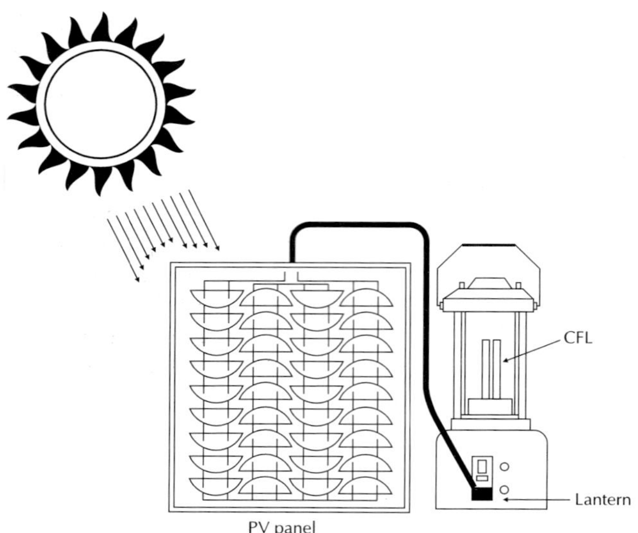

Figure 3 Schematic view of a solar lantern

Table 2 Solar lantern configurations as per MNRE specifications

Model	Lamp (CFL)	Battery capacity (at C/20 rate)	PV module rating
I-A	5 W	12 V, 7 Ah	8 to 9.9 W_p
I-B	5 W	12 V, 7 Ah	10 to 11.9
II-A	7 W	12 V, 7 Ah	10.00 to 11.9 W_p
II-B	7 W	12 V, 7 Ah	12.0 to 14.0

W – watt; Ah – ampere-hour; W_p – watt peak; CFL – compact fluorescent lamp
Source www.mnes.nic.in

As seen in Table 2, a 5-W or 7-W CFL is used, which can either be a two-pin type or a four-pin type. If a four-pin type CFL is used, then a preheating circuit is provided and the lamp is mounted on a base. The lumen output is generally in the range of 230±5% for a 5-W lamp and 370±5% for a 7-W lamp.

PV module

The PV module used in a solar lantern can have a power rating of 8, 10, or 12 W_p. Depending upon the latitude of the region, the module is installed at an optimum tilt angle on a rooftop or is placed where maximum sunshine is available. Normally, households do not install solar module on rooftops, but they are kept in the sun daily.

Battery

A sealed, maintenance-free, lead–acid battery or a tubular plate lead–acid battery with spillproof feature is used. The minimum battery capacity is of 6.5 Ah at 12 V or 12 Ah at 6 V.

Electronic components

An inverter with at least 80% efficiency is provided with the system. It also has a cut-off provision to prevent overcharging or deep discharging of the battery. A fuse is provided in the system to prevent short circuit. Like SHS, solar lanterns are also provided with two LED indicators to indicate the charging and the deep-discharge state. Other accessories include a 5-m-long wire extending from the module to the lantern, and an ON/OFF switch.

LED-based solar lantern

LEDs are now beginning to gain market acceptance primarily due to their reduced electricity consumption and extended lifespan. Solar power being expensive is aptly suited to the use of such LEDs. These LEDs appear in various colours but white LEDs are the most sought after by PV manufacturers. Quite recently, the state nodal agencies for renewable energy have begun to procure LED lanterns, along with the conventional lanterns using higher module power rating. The brief specifications of an LED solar lantern, as evolved by the MNRE, are described below.

Module

The module can be either crystalline silicon or based on thin film technology. The peak power rating of such a module should be more than 2.5 W at 17.0 V.

Battery

The LED-based solar lantern should have a sealed maintenance-free lead–acid battery with a capacity rating of 2.30 Ah at 12 V at C/20 discharge rate at 27 °C. Seventy-five per cent of the capacity should be between fully charged and load cut-off conditions.

Lamp

The lamp should be 5 mm, white LED (W-LED), diffused, and soothing to the eyes. The luminous performance of the LED must not be below 30 lumens per watt. Further, the light output from the source should be constant during the specified duration. The key technical specifications of an LED lantern are given in Table 3.

Table 3 Key technical specifications of LED lantern

Input voltage	12 V (nominal)
Input current	< 200 mA
Input power	< 2.5 W_p
Output power	> 2.0 W_p
Load disconnect battery voltage	11.20 ±0.1 V
Load reconnect battery voltage	12.50±0.1 V
High voltage cutoff	< 13.4 V
Duty cycle	minimum of 3–4 hours per day

V – volt; W_p – watt peak; mA – milliampere

Source PV company brochures

Outdoor lighting systems

The most common and widely used outdoor lighting system is the solar PV street light. Other applications include garden lighting, pathway lighting, and parking lights, which operate on the same principle.

PV street lights

PV street lights are stand-alone systems with one or two modules, a lead–acid battery, and a luminaire with a CFL. These are normally available in one model only, which is called the dusk-to-dawn system. It can be operated for around 12 hours. Key components of a solar street lighting system (Figure 4) are described briefly as under with electronic protections similar to those used in case of solar home systems.

Module

A single module of either 74 W_p or two modules of 37 W_p each are generally used. The operating voltage corresponding

PV street lights

to the power output mentioned above should be 16.4 V. The open circuit voltage of the PV modules under STC (standard temperature condition) should be at least 21.0 V.

Battery

The battery used is generally of flooded electrolyte type, positive tubular plate, low-maintenance lead–acid battery. It has a minimum rating of 12 V, 75 Ah (at C/10) discharge rate. For a battery to run smoothly, 75% of the rated capacity of the battery should be between fully charged and load cut-off conditions.

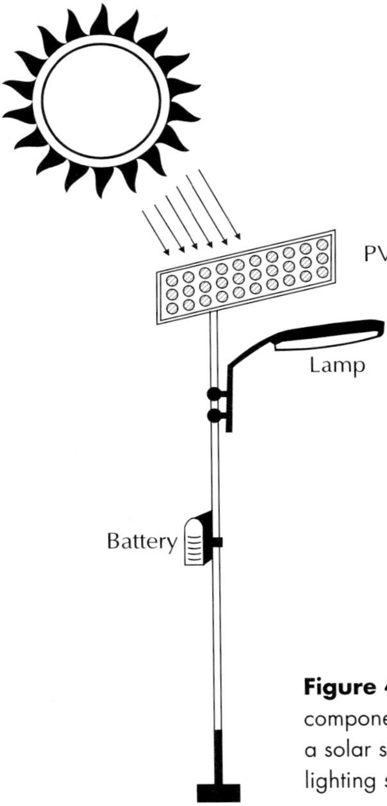

Lamp

These days, PV street lights use an 11-W CFL. It could either be of 2-pin type and 4-pin type. In the 4-pin type CFLs, preheating circuit is provided. The lamps are enclosed in a weather-proof assembly with a high reflecting base. The lamps are fixed in a base-up configuration. The output from the lamps is around 900±5% lumens.

PV panel

Lamp

Battery

Figure 4 Key components of a solar street lighting system

Electronic components

An inverter with a minimum 80% efficiency is provided with the system. It also includes a sensor for automatic switching (ON/OFF) from dusk to dawn (usually 12 hours). It also includes protection of the batteries from overcharging or deep discharging.

Other components

The PV module (s) and the lamp are mounted on a steel pipe at a height of 4 m above the ground. A metallic, corrosion-proof box is

provided to store the battery and a frame to mount
the modules.

Solar refrigeration

Developing countries like India need large-scale immunization
programmes, for which vaccines have to be kept within a specified
temperature range. The provision of refrigeration for this is
commonly known as the vaccine cold chain. Remote rural areas,
with little or no electricity at all, face one big problem in this
respect. Refrigerators run on kerosene or bottled gas do not
work well. Diesel-based systems face a problem of fuel supply, as
well as high maintenance. A good alternative in this scenario is
a solar PV refrigerator. Solar absorption (thermal) refrigerators
have fared poorly as against solar PV refrigerators. As per reliable
estimates, there are more than 6000 PV medical refrigeration
units in use worldwide. Vaccines are usually stored at temperatures
ranging between 0 °C and 8 °C. Figure 5 gives a schematic view of a
commonly used solar refrigerator.

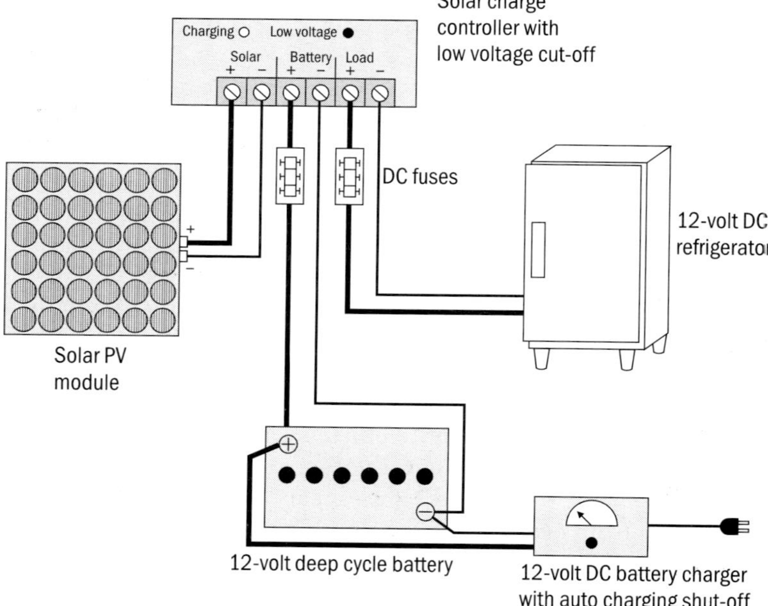

Figure 5 Schematic view of a commonly used solar refrigerator

Working principle

Solar PV refrigerators are based on the same principle as normal compression refrigerators. However, they use low voltage (12 V or 24 V) DC compressors and motors. Another major difference is that these have higher levels of insulation around the storage compartments to maximize energy efficiency. Most refrigerators have a freezer compartment for ice-pack freezing. Other systems have separate units for refrigeration and freezing. Usually available sizes range between 10 litres and 85 litres of vaccine storage capacity, with ice production rates of up to 6.4 kg per 24 hour.

In this case, the lead–acid deep cycle batteries are designed to run the refrigerators for up to five days, including no-sunshine days in a row. Here too, a charge controller performs the same functions as normally seen in case of lighting systems, for example. A roof or ground-mounted arrangement is used for PV array. The typical requirement is 150–200 W_p of PV modules. The energy consumption of a PV vaccine refrigerator is generally 400–800 Wh for a 100-litre refrigerator (without ice-pack freezing) and at +32 °C ambient temperature. However, if the ambient temperature is + 43 °C and freezing 2 kg of ice packs per 24 hour, the energy consumption of the same refrigerator increases to 900–1900 Wh per 24 hour.

Ideally, a good vaccine refrigerator must maintain correct internal temperature for a minimum of 10 hours when not connected to battery and solar array. In India, SPV vaccine refrigerators have been installed in more than 100 rural health centres. The capacity of these vaccine refrigerators is 50 litres (45 litres for vaccine storage and 5 litres for ice-making).

Technology advancement in solar refrigeration

The battery is often termed as a weak link in any PV system, besides being expensive. The WHO (World Health Organization) and UNICEF (United Nations Children's Fund) have recently supported a major partnership project titled 'Solar Chill'. These two international organizations have framed the performance standards for solar refrigerators. In this case, ice packs store energy in place of batteries. As of now, the solar chill models are undergoing extensive field trials. Table 4 summarizes the advantages and disadvantages of available vaccine refrigerator options tried so far.

Table 4 Advantages and disadvantages of available vaccine refrigerator options

Type of refrigerator	Advantages	Disadvantages	Remarks
Kerosene	Most common systems used for cold storage for vaccines	• Irregular fuel supplies • Poor fuel quality • Lack of effective temperature control • Constant operation and maintenance needs • Unreliable equipment	Large quantities of vaccines lost due to all these factors
Bottled gas	Usually more reliable than kerosene units	• Generally feasible in areas with assured supplies of bottled gas • Few operational problems still persist	
Solar PV	• Zero fuel cost • Low running costs • Lower maintenance needs • Higher performance reliability • Better temperature control • Longer equipment life	• Higher initial capital cost	Less quantity of vaccines is lost

Water-pumping systems

Groundwater is considered the safest and the most reliable amongst traditional sources of potable water. Conventional techniques for drawing groundwater for drinking and domestic use involve the use of the following.

- Mechanical energy provided by human beings and animals in conjunction with equipment—simple rope and bucket, handle or wheel with goatskin pouch, handpump, and footpumps
- Animal energy to turn flywheels made of wood or metal and use of diesel engine or electric motor-driven pumps

Table 5 presents a comparison between various water lifting/pumping technologies. There has been a significant increase in the use of renewable energy technologies for water pumping in the last few years, mainly due to constraints (cost, availability of fuels, and convenience) in the dissemination of conventional technologies.

Solar water-pumping system

Technologies using hydropower, solar energy, wind energy, and biomass are being used as alternatives. Each of these technologies offers certain advantages over others, but their viability is linked to many parameters specific to the location.

Table 5 Advantages and disadvantages of various water-pumping technologies

Technology	Advantages	Disadvantages
Hand pumps	Local manufacture possible Easy to maintain Low capital cost No fuel costs	Loss of human productivity Often inefficient use of bore-holes Only low flow rates achievable Maintenance of animals require feeding all-year round
Hydraulic pumps	Unattended operation Easy to maintain Low cost Long life	Require specific site conditions Low output High capital cost in some cases
Wind pumps	Unattended operation Easy maintenance Long life	Water storage required for low-wind periods Intensive project planning needs Difficult to install
Solar PV	Unattended operation Low maintenance Easy installation Long life	High capital cost Water storage required for cloudy periods Repairs require skilled technicians
Diesel pumps	Quick and easy to install Low capital cost Widely used Can be portable	Fuel supply erratic and expensive High maintenance cost Short life expectancy Noise and air pollution
Electric motor-driven pumpsets	Low capital cost Easy to install Wide range of applications	Electricity supply erratic Poor quality of supply in rural areas

Source IT Publications (1994)

PV water-pumping systems consist of a PV array, motor-pump unit, optional power conditioning equipment (for conversion of DC to AC, in case of AC pumps) (Figure 6). Two of the most common applications of PV water pumping are for irrigation and drinking water supply. A special project was initiated in India in 1984 to demonstrate and popularize the PV water-pumping system under the National Solar Photovoltaic Energy Demonstration Programme. These used a PV array capacity of around 360 W_p and a 0.5 HP (horsepower) motor-pump unit. The project covered the states of Andhra Pradesh, Bihar, Orissa, Tamil Nadu, Uttar Pradesh, and West Bengal.

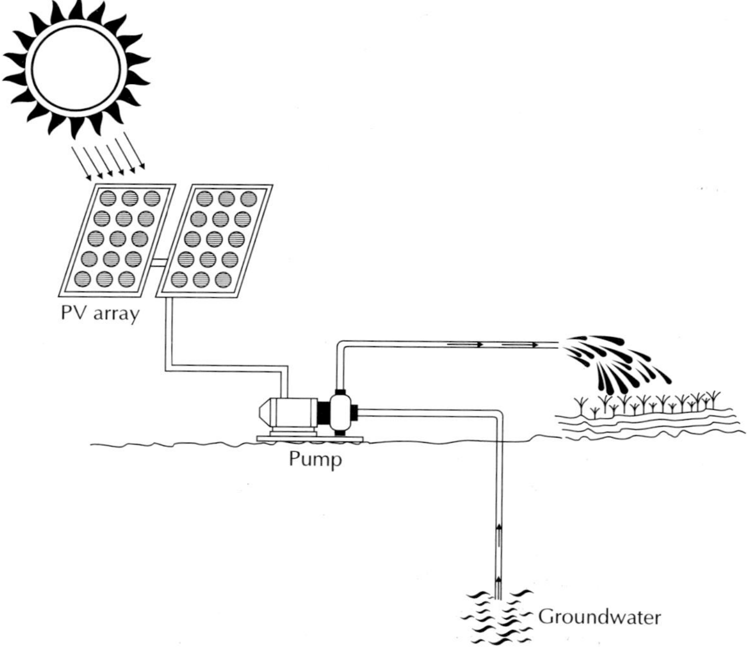

Figure 6 PV water-pumping system

The programme was coordinated by the then Department of Non-conventional Energy Sources, now known as the MNRE. By the end of December 2007, a total of 7068 PV water-pumping systems had been installed in the country. The capacity of a water-pumping system varies from 200 W_p to 2250 W_p. The PV water-pumping systems are being used mainly for drinking water, micro-irrigation, horticulture, and silviculture applications.

Irrigation

Another common application of PV pumpsets is for lifting water for irrigation. The following types of pumpsets are used for irrigation.

- DC submersible pumpset
- DC surface-mounted centrifugal pumpset
- DC floating pumpset
- AC submersible pumpset

The discharge from the pump at a suction head of 7 m or the total head of 10 m is approximately around 15 000 litres/day from a 200-W_p module and 170 000 litres/day from a 2250-W_p module. The discharge varies with the intensity of the sun, and is at its maximum during noon. At the total head of 10 m, a PV pump can irrigate about 0.5–0.6 ha (hectare), but this again depends upon the water table, soil conditions, and water management.

The cost of PV pumping installations varies, depending on the size, location, and other factors (such as labour and material costs). As with other PV systems, the main cost of water-pumping system is the initial capital cost. Clearly, greater the pumping head or daily volume of water required, greater is the energy demand—accordingly, the need for more modules. Modules form the major part of capital cost. However, once installed, very little maintenance is required. Hence, the annual recurring costs are lower.

Other applications

Apart from lighting, refrigeration, and water pumping, there are several applications for solar PV systems.

Battery-charging stations

In remote areas, there is a huge demand for supplying electric power to operate small torches (flashlights), radios, television sets, and emergency lights. This is evident from the continuous transportation of batteries from rural areas to grid-connected towns (for charging), battery-charging services offered by numerous DG (diesel generator) set operators in the villages and small towns, and generation of vast quantity of small 'throw-away' primary cells bought by the rural households. For example, in the Sundarbans, West Bengal, a family spends on an average Rs 12–15 per month on dry cell batteries.

Battery-charging stations consist of a PV generator, normally between 500 W_p and 1000 W_p, which charge the batteries centrally.

Solar battery-charging station

Batteries are either delivered or collected by local household owners. PV battery charging is very similar to SHS from a technical and economic point of view. Central charging stations are effective management systems for collecting fees or rentals, but the only drawback is the transportation of batteries (often perceived as a disadvantage by the users).

Box 1 TERI's PV charging station serves multiple functions

Dakshin Dimoria, a revenue village under Dimoria development block in Kamrup district, Assam, suffered from erratic power supply in spite of being electrified. More than 70% of households did not have access to grid connection, as they were too poor. For lighting purposes, the villagers mainly used kerosene lamps or paraffin candles, which not only provided insufficient lighting, but were also not environment friendly.

TERI launched the LaBL (Lighting a Billion Lives campaign) in Dakshin Dimoria in May 2008. The project was started in partnership with Kalang Jalachhed Samiti, an NGO involved with watershed development project in the area. Keeping in mind the villager's demand for other services in addition to lighting and their adaptability to the system, an STP (solar technology platform) has been created in the village. The STP was started in a small hamlet called Bahtala, which is accessed by five other hamlets in Dakshin Dimoria village. The STP consists of a solar-based charging station for charging 50 lanterns and a 100 Ah (ampere-hour) battery, connected through a Solverter™ developed by TERI, for powering other appliances such as computer, mobile phone, and water purifier. The STP is equipped to cater to the community's need for lighting, information and communication system, and other livelihood-related requirements. The villagers are eagerly waiting for more services from the charging station in the near future.

PV hybrid systems

PV systems generally have a provision for storing energy for a predefined period of insufficient sunshine. However, there may be exceptional periods of poor weather conditions when an alternative source is required to guarantee power production. In such cases, PV hybrid system is an appropriate solution. PV hybrid systems combine a PV generator with another power source – typically a diesel generator – but other renewable energy sources such as a wind turbine or hydro can also be combined (Figure 7).

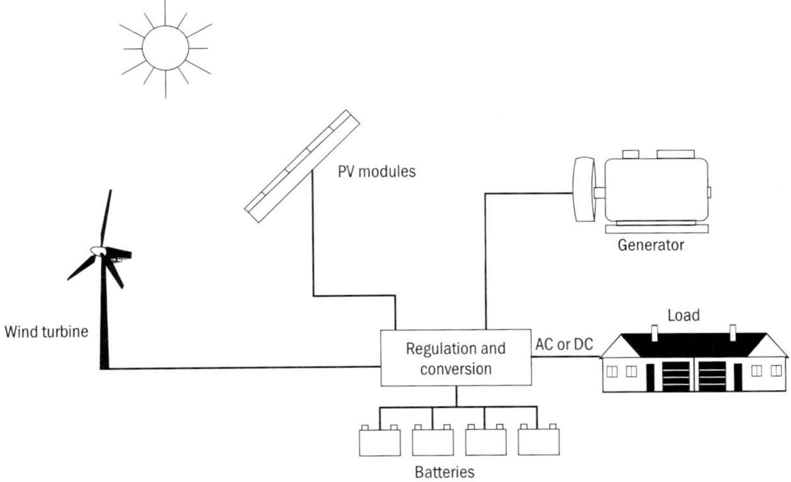

Figure 7 PV hybrid system

The PV hybrid system is usually sized to meet the base-load demand, with the alternative supply being used only when essential. Usually, the contribution of PV modules to the power generation is around 30% so as to keep the system cost down. This arrangement offers all the benefits of a PV system, such as low operation and maintenance costs, as well as a more secured energy supply.

PV hybrid systems are suitable in areas where occasional demand peaks are significantly higher than the base-load demand. It is not viable to size a system to be able to meet the demand entirely with PV if, for example, the normal load is only 10% of the peak demand. Likewise, a DG set sized to meet the peak demand would be operating at inefficient part-load for most of the time. In such a situation, a PV-diesel hybrid would be an ideal compromise.

PV teams up quite well with wind energy resource in a hybrid mode. For example, there is a 5-kW (kilowatt) PV–wind hybrid system in Chunnambar Island, in which the PV component is 1.8 kW and the wind component is 3.3 kW. Likewise, Minicoy, the southernmost island of Lakshadweep, has a 100-kW PV–diesel hybrid system in operation for the last few years.

Recently promoted urban applications

The MNRE has been encouraging the use of PV products and systems in rural areas for a long time. There is an increasing gap between the demand and supply of conventional power in urban areas. Taking a cue from this inadequacy of power, the MNRE has devised an incentive-based scheme for popularizing the following types of systems among potential customer segments in urban areas.

Street light solar control system

It is a common sight to see conventional street lighting systems glowing even in bright sunshine. A solar-based dusk-to-dawn automatic switching system takes care of switching off such lights at the specified timings. Thus, it leads to large savings in energy.

Solar-powered illuminated hoardings

Cities have large hoardings at important crossings. However, it is becoming difficult to keep these hoardings glowing due to paucity of power. Solar power can be used to run the small hoardings at least. A few state electricity regulatory commissions have already put a stop to the use of neon signs and illuminated hoardings.

Solar-powered traffic lights

Power failure in a traffic lighting system often leads to a messy traffic situation. Solar-powered LEDs are fully geared to direct the traffic during those periods in a stand-alone or a hybrid mode with conventional power

Solar road studs

A solar stud enables proper visibility on poorly lit roads at night. These can often be used between dusk and dawn and installed at vital points like zebra crossings and speed breakers.

Solar blinkers

It is a sign that is easily seen from a long distance. Solar-power-based LEDs are now being used ahead of any road-specific obstructions. The electronic controller enables constant current supply.

Solar traffic lights

Solar blinker

Building integrated PV

It is commonly known as PV for buildings or simply BIPV (building integrated PV). Solar modules can be put in place of walls or roofing panels within a building envelope. Any excess power generated is fed to the grid. It can also serve the purpose of peak load shaving.

Solar power packs

It is an annoying sight to see smoky and noisy generators at work in urban marketplaces. Solar power packs are ideally suited to replace such small generators run on petrol or diesel. Lights, fans, and computers can also be run on solar power, especially in emergency situations.

Reference

IT (Intermediate Technology) Publications. 1994
The Power Guide
London: IT Publications Ltd

Designing a photovoltaic system

Introduction

PV (photovoltaic) system design primarily depends on the type of application for which the power will be used. If power is required for pumping water for agriculture or drinking water supply, the electrical storage device is optional and not generally recommended because of added costs. When water pumping is not being carried out, the PV module capacity can be used for lighting application. However, for devices that are operated in the evening or at night, that is, for lighting, an electrical storage device is a must.

The overall design of a PV system, including the number of modules and batteries, depends upon the total system load. The load for a PV system is governed by the appliances to be powered by the system. Indoor lighting systems typically include lights, fans, and televisions, while outdoor loads include street lights, water pump, and so on. PV modules of varying peak watt ratings are available in the market. Batteries, which are measured in terms of ampere-hour (Ah), are also available in different capacities. The basic information required to design a suitable system for different applications is presented in Table 1.

Designing a PV system with battery back-up

In order to design a PV system with battery back-up, first the total load, or daily energy consumption, has to be determined. On the basis of existing or desired loads, the number of modules and the battery voltage required to meet the daily energy demand can be calculated. To ascertain the battery size and the required wattage of the modules, the peak load is calculated for such months when the maximum number of devices would be used. In addition, the months during which the intensity of solar radiation and the duration of sunshine hours are the lowest should be identified. In India, most of the systems are designed taking into consideration

Table 1 PV system application and associated information requirements

Application	Information required
Lighting and other domestic applications	Number of lights/appliances required
	Wattage of each light/appliance
	Average daily usage of each device in hours
	Seasonal loads and number of monsoon months
Solar pump	Water source (river, canal, well or borewell)
	Characteristics of source (diameter, water available, and so on)
	Head (total head from which water has to be lifted)
	Water demand (monthly basis and also peak season demands)
	Water storage and distribution
Solar refrigerator	Size of the refrigerating unit (volume)
	Maximum amount of goods to be frozen or kept
	Normal ambient temperature

the winter months, as well as an approximate number of rainy days during monsoon. A PV system is generally designed for the 'worst' month period.

Measuring solar radiation

The performance of PV systems is a direct function of available sunlight, or insolation, and the conversion efficiency of various devices. With increased insolation and improved device efficiencies, the system performance has increased. As the basic characteristics of any SPV module changes with the variations in solar radiation, the foremost parameter to be considered prior to installing any PV system is the availability of adequate solar radiation.

The daily solar energy that a PV array is capable of producing depends on the insolation observed throughout the day and the size of the array capturing the energy. In order to calculate the appropriate array size, one needs to know the availability of solar insolation at the site. For convenience, monthly solar radiation is usually expressed in terms of the daily average solar insolation of the month.

Insolation is measured in $kWh/m^2/day$. It ranges from an average of 3 $kWh/m^2/day$ to 7 $kWh/m^2/day$ worldwide. India receives solar energy equivalent to over 5000 trillion kWh, which is far more than the total energy consumption of the country. Under clear sky conditions, the daily average solar energy incidence is 4–7 kWh, depending upon the location in India. There are about 250–300 sunny days in most parts of the country. For example, the

Varanasi and Allahabad region receives an average summer solar radiation of 6.8 kWh/m^2 and an annual average of 5.8 kWh/m^2. It is only during winters that the average solar radiation falls to 4.8 kWh/m^2, and this occurs for less than 20% of the year.

Solar insolation is generally measured by a pyranometer. There are several types of pyranometers, ranging from analog to digitally read models. It is also possible to know the average value of solar radiation over a module surface in a given interval. Solar insolation can also be measured by a simple hand-held inexpensive meter known as 'Suryamapi'. It normally gives the value of solar insolation in milliwatt (mW/cm^2), which can easily be converted into a more acceptable unit of watt/m^2. Central Electronics Ltd was the first company in India to make that device using silicon solar cell as a sensing element.

Determination of load

To determine the load, a simplified procedure can be followed.

- Selection of units to be used
- Determination of wattage of each unit
- Determination of running hours of loads
- Determination of total load

Selection of unit

Depending upon the requirement, units – whether lighting or other – are identified first.

For example, let us assume that the total requirement of a household is as follows.

- Lights: three points
- TV (black and white): one point
- Fan: one point (during summer, four months)

Determination of wattage

The standard wattage for each electrical load is fixed. Table 2 provides the wattage of different electrical loads. In order to determine the system size, identification of the wattage of each electrical load is necessary.

For example, if CFLs (compact fluorescent lamps) are used for lighting in the example cited above, two CFLs of 11 W and one of 9 W are required. The required wattage for the household is given is Table 3.

Table 2 Wattage of different electric loads (AC/DC) for PV application

AC load	Wattage (W)
Fan	40/65
TV	
Black & white	18
Colour	40, 80, 120
Fluorescent tube	20, 40
CFL	
Solar lantern	5, 7
Domestic lighting	9, 11
Radio/tape recorder	5–20
Water pump	200–2250 W_p

DC load	Wattage (W)
Fan	20
TV	18, 20
Fluorescent tube	20

W_p – watt peak; CFL – compact fluorescent lamp

Table 3 Wattage requirement of sample household

Load	Number	Wattage (W)
CFL	2	11
CFL	1	9
TV (black & white)	1	18
Fan	1	20

CFL – compact fluorescent lamp

Determination of usage hours of loads

For each of the loads identified, the requirement in terms of usage hours per day is calculated. Table 4 shows the usage (hours) of each unit in the sample household.

Table 4 Usage hours of loads in sample households

Load	Number	Wattage (W)	Hours/day
CFL	2	11	5
CFL	1	9	3
TV	1	18	5
Fan	1	20	8

W_p – watt peak; CFL – compact fluorescent lamp

Total load determination

The total load is expressed in terms of watt hour/day (Wh/day). This is determined by multiplying the number of devices with their respective wattage and the number of hours used per day.

For example, for the case-specific household, a DC fan operational during summer (four months) and not in winter is taken into account. The system design is for maximum electrical load requirement. Total load calculated is shown in Table 5.

Table 5 Total load requirement of a sample household

Load	Number	Wattage (W)	Usage (hours/day)	Load (Wh)
CFL	2	11	5	110
CFL	1	9	3	27
TV	1	18	5	90
Fan	1	20	8	160
Total load				387

Hence, for this household, a PV system that can meet the daily energy demand of 387 Wh/day will be designed.

Number of PV modules required

The number of PV modules required to meet the daily energy demand can be calculated by either using the wattage of the module and the mean daily solar insolation in the region, or by first determining the array load and then the array size. The second method is the most frequently used one. System losses occur owing to use of battery, charge controller, and inverter. The efficiency of different system components is given below.

- Battery   80%
- Charge controller 90%
- Mismatch factor   0.85%

The average insolation during the least sunny days is 5.5 kWh/m^2/day in India. Using these values, one can determine the total array size or capacity of the system that will be required to meet the daily energy demand. Modules are available in certain rated capacities of 37, 40, 50, 74 W$_p$ (watt peak) and more. Accordingly, the number of modules that will be required to meet the total daily energy demand can be ascertained.

$$\text{Array load} = \frac{\text{Total daily load (Wh/day)}}{\text{Battery efficiency} \times \text{charge regulator efficiency}}$$

$$\text{Array size} = \frac{\text{Array load}}{\text{Insolation} \times \text{mismatch factor}}$$

Example For the total daily load demand of 387 Wh (of the sample household), the array load and the array size is as follows.

$$\text{Array load} = \frac{\text{Total daily load (Wh/day)}}{\text{Battery efficiency} \times \text{charge regulator efficiency}}$$

$$\text{Array load} = \frac{387}{0.8 \times 0.9}$$

$$\text{Array size} = \frac{\text{Array load}}{\text{Insolation} \times \text{mismatch factor}}$$

$$\text{Array size} = \frac{5.35}{5.2 \times 0.85}$$

$$= 129 \ W_p$$

Thus, modules that can deliver 129 W_p are required. The options available are to select one of the following, depending on cost and space available for installation.

- Four modules of 37 W_p (148 W_p)
- Three modules of 50 W_p (150 W_p)
- Two modules of 70 W_p (140 W_p)

Installation of a PV module

It is common practice to position a PV module facing south in the northern hemisphere and north in the southern hemisphere. Ideally, to gain maximum energy, the PV module can be mounted on a tracker to face the sun continuously, as this may increase the output up to 25%. As this is not practical in most cases, the best compromise for a fixed solar module is to face it square to the sun at noon. If the module is inclined from the horizontal surface of the ground at a 'tilt angle' corresponding to the latitude of that location, then it will be square to the sun at noon at the equinoxes (21 March and 21 September). A steeper angle increases the output in winter, while a shallower angle gives more output in summer.

In practice, it is preferred that the module should be fixed at an angle corresponding to the latitude of the place, to which it is necessary to either add or subtract another 10 degrees, depending on the season. In winters, 10 degrees are added to the latitude, while in summers, 10 degrees are subtracted so as to receive maximum solar insolation. In the equatorial and tropical regions, the tilt angle should be at least 10 degrees in order to ensure good run-off (of rainwater) to keep the modules clean and also for better cooling. Table 6 gives the optimum tilt angle at different latitudes. Concentrator modules are usually tracked both manually, as well as a through micro-processor-based control mechanism. This is because such modules utilize only the direct light and not the diffused form of light.

Table 6 PV module tilt angle

Latitude (degree)	*Optimum tilt angle* (degree)
9	15
10–20	latitude +5
21–45	latitude +10
46–65	latitude +15
66–75	80

Source Sandia National Laboratories (1996)

Determining the battery bank

Most of the PV systems need batteries to run loads during non-sunshine hours. The battery rating is given in terms of Ah (ampere-hour) and rating. Therefore, a 12-V, 40-Ah battery also has a 10-hour discharge rate mentioned with it. This means that a 12-V battery would have 480 Wh (watt-hour) (12 V × 40 Ah) and the power can be drawn at the rate of 4.8 A (ampere) over 10 hours.

Reserve day

Reserve day is a term used for indicating the minimum back-up energy that is required to be stored in the battery. It is determined according to the prevalent climatic conditions. In India, in the north-eastern states and the Himalayan belt (areas receiving very high rainfall), a back-up of three days is generally provided with the system. In the remaining parts, two days of reserve are sufficient. Reserve day is also known as the autonomy of a PV system. It generally ranges between one and five days, depending upon the criticality of a PV application, for example, a telecom equipment deployed atop a hill will require more reserve days.

Battery bank and battery number

Battery sizing is one of the most critical factors of system sizing. It depends largely on the region where the system is being installed. Configurations of batteries commonly used in PV systems in India are given in Table 7. While designing the battery capacity, it is important to keep in mind that the batteries cannot be fully discharged. As a safety measure, care is taken to see that the battery is not discharged beyond 50% of its capacity, as the battery cannot be charged completely once this state is reached and will always operate below its rated performance. In some calculations, this value is taken as 20%.

Table 7 Configuration of batteries commonly used in PV systems

Nominal voltage (V)	*Ampere-hour* (Ah)
12	7
12	40
12	70/75
12	100

$$\text{Battery capacity} = \frac{\text{Daily load (Wh)} \times \text{reserve back-up}}{\text{Nominal voltage} \times \text{maximum allowable discharge}}$$

For example, if the daily load is 387 Wh (sample household), then the number of batteries required for a system with two days' reserve back-up will be as calculated below.

Load	=	387 Wh
Reserve days	=	2
Maximum allowable discharge	=	50%

$$\text{Battery capacity} = \frac{\text{Daily load} \times \text{reserve days}}{\text{Nominal voltage} \times \text{maximum allowable discharge}}$$

$$\text{Battery capacity} = \frac{387 \text{ Wh/day} \times 2 \text{ days}}{12 \times 0.6}$$

$$= 744 \text{ Wh/6}$$

Battery capacity = 129 Ah

To determine the number of batteries, we can use the above formula and decide on the combination. Either four batteries of 40 Ah or two batteries of 70 Ah can be selected.

System design

The system design includes the number of modules, batteries, and the load points. Besides these, other components include charge controller, inverter, and mounting equipment. The system design for the sample household would be as follows.

Modules	: $4 \times 37 \text{ W}_\text{p}$
Battery	: $4 \times 40 \text{ Ah}/2 \times 70 \text{ Ah}$
Load	: $2 \times 11 \text{ W CFL}$
	$1 \times 9 \text{ W CFL}$
	$1 \times 18 \text{ W b/w TV}$
	$1 \times 20 \text{ W DC fan (summer months)}$

Reserve back-up : 2 days

The number of modules required will also depend upon the place of installation and the available space. The same applies with the batteries. Another factor that will influence the number of modules and batteries is their market price.

Designing a PV system without battery back-up

The most common application of this type of system is the solar water pump, used for supplying drinking water or for irrigation. The procedure for designing such systems involves ascertaining the daily water requirement, which determines the load and size of the PV array. It is important to note that the requirement of water either for irrigation or for drinking water (water stored in a tank) is mainly during the period when adequate solar insolation is available. So, there is no need for an electrical storage device (battery). However, when there is little or no demand for water, the available PV capacity can be utilized for lighting applications at the same location. Also, it is not economically feasible to have a battery bank in case of a water-pumping system.

Determining the total water head

This is the most important step in designing water-pumping systems. The head (in metres) indicates the level from which the water has to be lifted, or pumped. Both static head and dynamic head have to be calculated. Static head is the height over which the water has to be pumped, and dynamic head is essentially a function of friction losses and the drawdown. In case of PV pumps for irrigation, the total water head is the sum of static head and dynamic head. In case of pumps for drinking water supply, the total head is the sum of the total head (static and dynamic) and the height of the storage tank from the ground level.

Total water head = static head + dynamic head (PV irrigation pumps)

Total water head = static head + dynamic head + height of storage tank

Example If a PV pump for drinking water is to be installed in a village, then water is to be pumped from a height of 15 m. The water then has to be pumped to a storage tank that is at a height of 15 m. Dynamic head of the system is approximately 7 m (frictional losses of lifting water to the storage tank are included).

Total water head = Static head + dynamic head + height of storage tank
= 15 + 7 + 15 m = 37 m

Determining the water demand

Water demand is the minimum quantity of water that is required by a person. According to the Government of India standards, a minimum of 40 litres of water per person per day should be made available. The water demand is always calculated for the targeted year. The population of the village is projected for the design period. In case of PV pumps for agriculture-related activities, the daily water requirement in terms of cubic metre is taken into consideration.

For example, if the population of the village in 1998 was 500 (100 households), and the population in 2008 (the design period) is 625, then the daily water demand of the village in the year 2008 is to be calculated. In some calculations, the per capita increase in water requirement of a person is also included. In this calculation, we assume that the per capita water demand would remain constant at 40 litres per day.

$$\text{Daily water demand} = \text{Population} \times \text{per person daily water demand}$$
$$= 625 \times 40 = 25\,000 \text{ litres/day}$$
$$= 25 \text{ m}^3/\text{day} \ (1 \text{ m}^3 = 1000 \text{ litres})$$

Determining the system size

The system size in case of drinking water supply would also include the number of days for which the water has to be stored. Generally, two to three days of water storage is accounted for in designing the system. However, the first step is to determine the total energy that is required to lift the water. Although there is no charge controller or battery involved in the system, certain conversion losses at the time of water pumping occur. The average solar insolation in India is 5.5 kWh/m^2/day.

$$\text{Energy required} = \text{mass of water} \times \text{gravity} \times \text{total water head (in joules or kilojoules [kJ])}$$

For example, if the volume of water that has to be pumped for the design year of 2008 is 25 m^3/day and the total head is 37 m, the inputs will be as follows.

$$\text{Energy required} = 25 \times 9.8 \times 37 \times 0.28$$
$$= 2538 \text{ Wh/day} \ (1 \text{ kJ} = 0.28 \text{ Wh})$$

$$\text{Array load} = \frac{\text{Energy required}}{\text{System efficiency}}$$

$$= \frac{2538}{0.85} = 2986 \text{ Wh}$$

$$\text{Array load} = \frac{\text{Array load} \times \text{reserve days}}{\text{Insolation} \times \text{mismatch factor}}$$

$$\text{Array load} = \frac{2986 \times 2}{5.5 \times 0.85}$$

$$= 1277.4 \ W_p \ (\text{say } 1278)$$

System design

The PV system for the case-specific village would require a PV array of 1278 W_p to lift water from a height of 15 m to an overhead tank (at a height of 15 m above ground) and use the water for distribution. Yet another important aspect in designing the system is to determine the type of water pump (DC pump–submersible or surface mounted) that is available in the market. Depending upon the cost of modules, one can decide on the capacity of the module to be used. Most of the PV pumps in India use 74 W_p modules. Thus, in the village, if 74 W_p modules are used, then the number of PV modules required will be around 18.

Finally, let us take a look at the criteria that can lead us to choose a quality PV system. The idea is to derive the best possible performance from a system under actual field operating conditions.

Criteria for a good quality PV system

The following are the most important considerations deserving a system designer's attention.

- The system should be properly sized and oriented to provide the expected electrical power and energy.
- It should use sunlight- and weather-resistant materials for all outdoor equipment.
- It should be installed with minimum of shading from objects like foliage, vent pipes, or adjacent structures.
- It should be installed in compliance with all applicable building and electrical codes.
- It should be installed with a minimum of electrical losses due to wiring, over-current protection, switches, and inverters.
- It should be properly grounded to reduce the threat of shock hazards and induced surges.

System simulation procedures

The size of a PV system can be easily calculated in the manner presented above. However, such a procedure does not take into account all the site-specific parameters and a few others like financial analysis. A more complete procedure involves the use of computer-enabled techniques based on well-established tools. These tools are more commonly known as the system simulation software. Free versions of these software are normally available on the respective websites of the developers. The only thing that you need at your end is a computer with an Internet connection. The following are some of the well-recognized software in terms of their actual usage.

RETSCREEN

It is one of the most widely used tools used not only for PV system sizing, but for a whole range of renewable energy technologies. Around a 100 professionals across the world have contributed to its development. It is now regarded as a unique decision support tool. RETSCREEN is widely used to evaluate the following parameters, which are of significance to a system designer.

- Energy production
- LCC (life cycle costs)
- GHG (greenhouse gas) emission reductions

A unique feature of the software is that it incorporates product and weather databases, including that for India. There are five important worksheets in MS Excel format with each technology workbook file meeting the following purpose(s).

- Calculate the annual energy production of the RET project
- Equipment data worksheet
- Cost analysis worksheet
- Optional GHG emission worksheet
- Financial summary worksheet

There are many financial indicators like the IRR (internal rate of return), NPV (net present value), and the simple payback. The conclusive summary is then available in the simple project cash flow graph for the twin purpose of ready referencing and decision making. It also enables a tax analysis and a CDM (Clean Development Mechanism) analysis in case of individual projects.

You may download the latest version of this freely available software – RETSCREEN – from <www.retscreen.org>.

PV-Watts

PV-Watts is a powerful energy simulation software designed by the NREL (National Renewable Energy Laboratory) of the US

Department of Energy. It calculates the energy produced by a grid-connected PV system. Within a grid-connected system, PV modules are joined together to form a PV array. DC electricity produced by the modules passes through an inverter to change it into AC power. In case the PV system produces more power than the system user needs, the inverter sends the excess power to the utility grid. In the reverse situation, when the user needs more power than what is being produced, the power utility supplies AC power to the user in a desired measure. This inflow and outflow of energy is measured through what is commonly known as the net metering. PV-Watts takes care of all these situations and tells a potential user beforehand about the expected performance from a PV-grid interfaced system.

You may download the latest version of this freely available software – PV-Watts – <from www.pv-watts.com>.

HOMER

This is basically a computer model that does the task of working out various system design options for both the off-grid and grid-connected power systems for remote stand-alone and distributed generation applications. HOMER's optimization and sensitivity analysis algorithm gives you an extra choice to evaluate the techno-economic feasibility of a large number of technical options. It also accounts for the variations in technology costs and energy resource availability.

You may download the latest version of this freely available software – HOMER – from <www.homer.com>.

There are several other computer-aided programmes like NSOL, which can be used to work out the size of any PV system.

Conclusion

System designing depends upon the application and the type of loads to be operated. The load for solar home lighting systems is low, so smaller systems are used. For drinking water supply or irrigation needs, the demand and the consequent load are much higher. In such cases, the system design aims to optimize the available sunshine hours, and hence configurations are of higher capacity (large systems). A similar process can be followed for system design in case of other applications such as PV refrigeration. The chapter has provided an overview of system designing. Most of the manufacturers/system installers normally customize the system design for the users. However, systems based on the MNRE specifications are normally available in the pre-designed form as well.

The computer-based system sizing procedures provide a more detailed insight into the expected functioning of various types of PV systems in accordance with the variation in site specifics, user profile, end-use requirements, and importantly, the daily hours of use. It is quite possible to work out the financial viability of the entire system (under consideration), taking into account many critical parameters like the annual energy generation, life of the system, and resource availability. This procedure is usually known by the name of cash flow statements and is of great help to arrive at the techno-economic viability of any PV application.

Reference

Sandia National Laboratories. 1996
Stand-Alone Photovoltaic Systems: a handbook of recommended design
Details available at <http://www.sandia.gov/pv/docs/PDF/ Stand%20Alone.pdf>, last accessed on 8 September 2008

Cost of photovoltaic systems

Introduction

The economic viability of a PV (photovoltaic) system is largely
dependent on the end-use application and site specifics. The cost
of PV systems can be highly variable, depending upon location,
source, volume delivered, quality of systems, and availability
of financial and fiscal incentives (taxes, subsidies or loan rates
applicable). Generally, PV systems are not economically viable on a
direct market cost basis. For instance, the cost of a PV array is more
than the cost of a DG (diesel generator) set. However, the life of a
DG set is 10 years, whereas the life of a PV module is expected to
be around 25 years—more than double the life of a DG set. Table 1
gives the expected lifespan of different PV components.

Table 1 Typical lifespan of PV components*

Component	Years
Module	25
Charge controller	15
Inverter	5–15
Batteries (solar)	4–6
Wiring	> 10

* As per manufacturers' ratings

The discounted LCC (life cycle cost) analysis method is
often used to explain the economic viability of PV systems when
compared to a DG. The price of a PV module fluctuates, largely
due to shortages of silicon in the international market. It also varies
from one manufacturer to the other. In 1991/92, the price per
peak watt of a module was Rs 225, which has gradually come down
to Rs 195 in 2007/08. It is expected to come down further as new
production capacities based on the most modern techniques take
shape, along with expanding markets.

According to the LCC analysis method, PV systems are much more economically viable than alternative sources such as grid and diesel if the loads are dispersed and smaller. This is because in case of smaller loads, the system can be designed accordingly to meet the demand. However, in case of DG sets, one has to buy a minimum practical sized generator, which may still be oversized for rural applications. The cost of PV systems would decrease further if there is an increase in the volume of production. Lately, new production capacities are being established with capacities in the range of 80 MW (megawatt) or more. The production of modules has been increasing steadily to meet the existing demand. Over the last few years, the average annual growth of in-house production of modules has been 20%–25%.

Nearly 45 MW_p of solar cells and 80 MW_p of solar modules were produced locally during 2006/07. Table 2 gives the annual production of solar cells and modules in India.

Table 2 Annual production of solar cells and modules in India

Period	Solar cell production (MW_p)	Solar module production (MW_p)
1985/86	0.5	0.5
1990/91	1.0	1.0
1995/96	4.2	8.0
2000/01	14.0	17.0
2001/02	20.0	20.0
2002/03	22.0	23.0
2003/04	25.0	36.0
2004/05	32.0	45.0
2005/06	37.0	65.0
2006/07	45.0	80.0

MW_p – megawatt peak
Source MNRE (2007)

Factors that influence the purchase of PV systems

It is important to understand that the customers do not buy PV systems based on LCC analysis. The major factors are that influence the purchase of PV systems are listed below.

- Need
- Successful demonstration (opinions gathered from various quarters)
- Upfront cost
- Availability of subsidy/concessional loan facilities
- Ease of O&M (operation and maintenance)

- Reliability and confidence in technology
- Availability of spare parts and repair and maintenance facilities

The high upfront cost of PV systems is still one of the major constraints in its large-scale dissemination. The market price of the systems varies from manufacturer to manufacturer in case of direct selling as against the lowest tendered price for government-administered programmes. Nearly all cell and module manufacturers like Central Electronics Ltd, Udhaya Semiconductors, Bharat Heavy Electricals Ltd, Tata–BP Solar, and Rajasthan Electronics and Instruments Ltd market complete systems. In addition to these, there are several system integrators in the market, who buy components (like solar modules and batteries) of PV systems from big manufacturers, and then assemble and market them.

Subsidies and soft loans

The purchasing decision of the consumer is greatly influenced by the tax or subsidies that the government provides on these systems. The MNRE (Ministry of New and Renewable Energy), Government of India, provides subsidy on the system cost. Depending upon the application, the subsidy varies. Of late, the subsidy structure has been revamped by the concerned ministry with a clear distinction between the rural and urban uses as well. In addition to this, IREDA (Indian Renewable Energy Development Agency) provides soft (low-interest) loans to the users of the system.

Working out the cost

PV system costing is done using the per watt peak cost of the module. There are varying estimates available on the cost contribution of the solar modules, ranging from 42% to 52%. The thumb rule is that the total cost of the PV module is nearly 50% of the system cost. In the BoS (balance of system), the battery cost is nearly 15%, and the cost of the charge controllers, inverters, and other electrical equipment is also 15%. The balance 20% consists of the indirect costs, which mainly includes taxes. It is logical to arrive at the best system cost estimates after evaluating a suitable system design. Once the system designing has been completed, one can calculate the approximate cost of the system.

For example, based on the system design with back-up, as mentioned in the previous chapter, the expected expenditure on the system can be calculated. The system design for the household was a PV system with total power requirement of 129 W_p. If the per watt peak cost of the module in 2007 is considered, the approximate cost of the PV system can be worked out. In 2007, the cost per peak watt was Rs 195. Therefore, the cost of the modules is approximately Rs 25 155. If the thumb rule of 50% is taken into consideration, then the system cost becomes Rs 50 310.

Pricing estimates of PV products in the Indian market

The manufacturing cost of solar cells and modules is coming down due to several reasons.

- Wafers now use reduced thickness of expensive silicon material.
- Solar-to-electric conversion efficiency of cells is steadily improving.
- New production capacities are being set up on a much higher scale of magnitude.

Table 3 shows the indicative prices of the most commonly used solar products and systems. It may be mentioned here that the quantum of state subsidy for such products varies from one place to the other. However, the central subsidy is more or less fixed in accordance with the backwardness of a particular region and end user categories.

Fiscal incentives

- No industrial clearances are needed for setting up a PV industry.
- No clearance is needed from the CEA (Central Electricity Authority) for generation projects of up to Rs 1 billion.
- Customs duty concession is available for the spares and equipment.
- Excise duty on a number of capital goods has either been reduced or exempted.
- Soft loans are available through IREDA.
- Financial support is available for the PV industry for R&D (research and development) projects in association with technical institutions.

Foreign direct investment

- Foreign investors can enter into a JV (joint venture) with an Indian partner for financial and/or technical collaboration.

Table 3 Indicative prices of commonly used solar products and systems

Product/system	Brief description	Total cost (Rs)	Central subsidy	Delivery mechanism
Solar lantern	Module 10 W_p CFL 1x 7 W battery 12 V, 7 Ah	3 300	Rural application (subsidy varies)	State nodal agencies
Home lighting system Model-I	Module 18 W_p CFL 1x 9 W battery 12 V, 20 Ah	7 177	Rural application (subsidy varies)	State nodal agencies
Model-II	Module 37 W_p CFL 2x 9 W battery 12 V, 40 Ah	12 750		
Model-III	Module 37 W_p CFL 1x9 W, 1x20 W (12" DC fan) battery 12 V, 40 Ah	12 850		
Street lighting system	Module 74 W_p CFL 1x18 W battery 12 V, 70 Ah	24 000	12 000 or 50%, whichever is less	Municipalities
Street light controller	Module 5 W_p	25 800 (including AMC for 5 years)	5 000 or 50%, whichever is less	Municipalities
Road stud		1500	1 000 or 50%, whichever is less	Municipalities/police
Road blinker	Module 37 W_p	19 000	7 500 or 50%, whichever is less	Municipalities/police
Power pack	Array 1 kW_p	359 000	100 000 or 50%, whichever is less	Municipalities
Water pumping system	Array 1800 W_p motor 2 HP	290 000	180 000	All types of users

W_p – watt peak; CFL – compact fluorescent lamp; V – volt; Ah – ampere-hour; AMC – annual maintenance contract; kW_p – kilowatt peak; HP – horsepower
Source Compiled from various sources

- Proposals for up to 100% foreign participation in a JV qualify for an automatic approval.
- Government encourages foreign investors to set up projects on BOO (build, own and operate) basis.

Conclusion

With the initial capital cost of a PV system still high, it is essential that adequate credit facilities are made available to the end users in accordance with their paying capacity. Over the last few years, the ministry has been trying to reduce the quantum of subsidy for a variety of end-use applications. The immediate purpose is to stimulate the development of a non-subsidy market in a phased manner. A right balance of subsidy and low-interest loan is still needed to promote the PV systems on a wider scale. New production capacities are being set up, which may lead to lower costs in the near future. The fact is that there is still a large market waiting to be tapped, particularly for lighting applications, across all geographical regions in the country.

Reference

MNRE (Ministry of New and Renewable Energy). 2007
25 Years of Renewable Energy in India
New Delhi: MNRE, Government of India

Maintenance of photovoltaic systems

Introduction

Any solar PV (photovoltaic) system is not absolutely maintenance-free. A minimum level of maintenance is needed to keep it running optimally. A PV system may fail in the absence of proper repair and maintenance facilities. Maintenance ensures that systems operate efficiently; it also prevents the occurrence of problems in future.

For the last few years, the MNRE (Ministry of New and Renewable Energy) has accorded due importance to the need for system maintenance. It has been earmarking part of the programme funds as service charges payable to the system supplier on a per unit basis. This is a marked departure from the early days of having a fixed-term maintenance contract with those supplying the systems. The system procurement agency in most of the government-supported programmes happens to be the state nodal agency for renewable energy.

Maintenance of PV systems requires minimal time, involving a simple procedure. Minor steps can be taken care of by the owners themselves. However, major maintenance needs require trained technicians, who are familiar with the PV systems and associated components, and who know the proper safety procedures. The basic information on the maintenance procedures, dos and don'ts of PV systems, as well as troubleshooting tips, are provided later in this chapter.

Maintenance procedures

Maintenance of a PV system involves a basic physical examination of several component categories. Since some PV systems (large capacity systems) operate at high voltages and currents, proper safety procedures must be followed. The use of appropriate tools and materials is required for proper maintenance. The general procedure for maintenance of the systems is as follows.

- Complete physical inspection of the system at least twice a year
- Developing and maintaining inspection forms and records

PV module

The PV module is one of the most vital components of a system, as it is the source of electricity generation. Modern day modules have a life of more than 20 years, and are generally problem-free. The most common problem in the modules is that of insufficient electricity generation. This may be either due to accumulation of dirt, shading on the modules, electrical problems, incorrect orientation or tilt angle. The maintenance procedure involves the following.

- Check for delamination, discolouration, cracked or broken cells, excessive dirt, or bird droppings.
- Check for shadows and shading in early mornings and during late afternoons.
- Ensure that there is no excessive vegetation growth around the module.
- Ensure that modules are well attached to the frame, and check the frame for corrosion.
- Inspect junction box seals for cracks and moisture.
- Make seasonal tilt adjustments as required.
- Measure current and voltage of individual modules/array.

At the time of installing the PV system, it must be ensured that there is no shading from the modules (if more than one module is used) on each other. Some of the common dos and don'ts are mentioned below.

Dos

- Clean the module surface regularly with a clean, dry/wet cloth
- Check all the cable and wiring connections for their firmness.
- Check that there is no corrosion in the modules and there is no seepage of rain water.

Don'ts

- Do not use any detergent for cleaning the module.
- Do not use sharp-edged materials to remove bird dropping.
- Do not touch the surface of the module with oily or greasy hands.
- Do not attempt to fiddle with the circuits and the wiring.

It is essential to remember that there is very little that can go wrong with the modules. However, the solar cells in the module are so fragile that once dismantled, putting them back is almost like installing a new module. So care must be taken to maintain the module.

Batteries

One of the main problems highlighted in evaluation studies related to systems was the failure of batteries. Users often assume that the battery has sufficient storage and, therefore, loads can be operated for a longer duration. The manufacturers and the system installers often design a system in which the battery would reach a discharge stage if the current withdrawn from it is more than 50%. The following procedure should be followed for maintenance of batteries.

- Inspect battery terminals for corrosion and loose cables.
- Check battery surface for electrolyte leakage.
- Ensure that batteries are not in direct contact with the floor.
- Check battery enclosure box for proper ventilation.
- Check the level of electrolyte and distilled water in the batteries; top up the battery if the level is below the specified mark.
- Check the specific gravity of the cells at regular intervals.

The battery must be placed in an enclosed box. It should be kept away from direct human contact. Some of the common dos and don'ts are given below.

Dos

- Charge the battery every day.
- Clean the battery top with a cloth/brush with baking soda and water solution.

Don'ts

- Do not keep any inflammable objects near the battery.
- Do not keep the battery exposed; always keep it in a sealed box that has proper ventilation.
- Do not put additional load on the battery, as it will reduce its life.
- Do not drop any metallic objects on the battery terminals—it may lead to sparking.

It is essential to remember that even at a very low voltage, a large battery bank can release high current. So, it is always safe to wear protective gear at the time of battery maintenance, as there is a possibility of explosion.

Electronic components

Field performance evaluation studies have indicated that solar lanterns occasionally failed due to problems in electronic components. It is essential, therefore, to regularly maintain the electronic components as given below.

- Verify voltage set points of charge controller with regard to battery specifications and system requirements.
- Check connections for corrosion and loose wires.
- Check for any unusual noises from charge controller.
- Ensure that charge controller is in a sheltered, clean, and well-ventilated environment.

Wiring

- Check wiring connections for corrosion and loose connections.
- Check wire insulation for degradation.
- Ensure that all metal equipment cases and frames in the system are well grounded (all the way to the grounding object).
- Check the operation of all LEDs (light emitting diodes)
- Check circuit breakers and fuses.

These are some of the common maintenance procedures that can ensure long life of the PV systems. Some of the problems in the systems can be easily detected and corrected. Tables 1–6 on troubleshooting outline actions that need to be initiated if problems arise. The following is a list of commonly used tools/material for system maintenance/trouble-shooting.

- Multimeter
- Hydrometer
- Ampere-hour meter
- Pliers, fuses, wires and connecting pins
- Distilled water
- Spares

The most commonly observed problems vis-à-vis PV system installations are summarized below.

- Semi-trained or untrained personnel entrusted to install the system
- Improper attachment of PV array to ground
- Poor cleaning of the module surfaces
- Unsafe installation of batteries
- Irregular use of batteries
- Poor maintenance (in terms of topping up or other aspect)

Table 1 Troubleshooting for PV panels

Problem	Cause	Result	Action
No current from array	Switches, fuses, or circuit breakers open, blown, or tripped; or wiring broken or corroded	No current can flow from array	Close switches, replace fuses, reset circuit breakers, repair or replace damaged wiring
Array current low	Some modules shaded	—	Remove source of shading
	Some modules damaged or defective	—	Replace the module
	Full sun not available	—	Remove source of shading/check the orientation of panel
	Modules dirty	—	Clean the module surface
	Array tilt or orientation incorrect	—	Correct the tilt angle

Source NFEC (1989)

Maintenance of photovoltaic systems

Table 2 Troubleshooting for batteries

Problem	Cause	Result	Action
No apparent battery defect but low battery charging	Load too large, switched on for too long, or inadequate sun	Battery is always in a low state of charge	Reduce load size or increase system size
	Batteries too cold	A higher voltage is required to reach full charge	Insulate battery enclosure
Low electrolyte level	Overcharging	Loss of battery capacity	Add distilled water, unless batteries are damaged beyond repair
Voltage loss overnight even when no loads are ON	Faulty blocking diode	Reverse current flow at night, discharging batteries	Replace diode
Voltage not increasing even when no loads are ON and the system is charging	Faulty charge controller	No power from array going into batteries	Repair or replace charge controller
	Loose, corroded, or broken wiring	Less power from array going into batteries	Repair or replace damaged wire
	Shaded modules, broken cell, or disoriented panels	Array output reduced	Remove source of shading, replace module, or correct module orientation
	Wiring too long or undersized	Voltage reduced	Check wire size
No apparent battery defect	Load too large, ON too long, or inadequate sun	Battery is always in a low state of charge	Reduce size of load or increase size of system
High water loss	Overcharging	Heat damage to plates and separators	Replace battery, repair or replace charge controller

Source NFEC (1989)

Table 3 Troubleshooting for charge controllers

Problem	Cause	Result	Action
Controller not charging batteries	Poor connection at battery terminal	Charge controller finds batteries cooler than their actual temperature	Repair, readjust, or replace charge controller
Battery voltage loss overnight even when no load is ON	Faulty blocking diode, no diode or faulty charge controller	Reverse current flow at night, discharging batteries	Replace or add diode, or repair or replace series relay charge controller
	Old or faulty batteries	Batteries self discharging	Replace batteries
Fuse to array blows	Array short-circuited with batteries still connected	Too much current through charge controller	Disconnect batteries when testing array's short-circuit current
	Current output of array too high for charge controller	Too much current through charge controller	Replace charge controller with a higher rating one
Fuse to load blows	Short circuit in load	Unlimited current	Repair short circuit or reduce load size
	Current draw of load too high for charge controller	Too much current through charge controller	Reduce load size or increase charge controller size
	Surge current draw of load too high for charge controller	Too much current through charge controller	Reduce load size or increase charge controller size

Source NFEC (1989)

Maintenance of photovoltaic systems

Table 4 Troubleshooting for inverters

Problem	Cause	Result	Action
No output from inverter	Switch, fuse, or circuit breaker open, blown, or tripped, or wiring broken or corroded	No power can move through inverter	Close switch, replace or reset fuse or circuit breaker, or repair wiring or connections
	Low voltage disconnect on inverter or charge controller circuit is open	No power available to inverter	Allow batteries to recharge
	High battery voltage	Inverter does not start	Connect load to batteries and operate it long enough to bring down battery voltage

Source NFEC (1989)

Table 5 Troubleshooting for system wiring, switches, and fuses

Problem	Cause	Result	Action
Load does not operate at all	Switches in the system are turned OFF or are in the wrong position	PV electricity cannot be supplied to loads or batteries	Put all switches in correct position
Load operates poorly or not at all	There is a high voltage drop in the system. Check for under- or over-sized wiring, over-sized loads, or a defective diode	Inadequate voltage to charge batteries or operate loads	Increase wire size, reduce load size
	Wiring or connections are loose, broken, burned, or corroded	—	Repair or replace damaged wiring or connections
	Wiring or connections are short-circuited or have a ground fault	—	Repair short circuit or ground faults

Source NFEC (1989)

Table 6 Troubleshooting for loads

Problem	Cause	Result	Action
Load does not operate at all	Load is too large for the system or inadequate sunshine	Shortened battery life, possible damage to loads	Reduce load size or increase array or battery size
	The load is in poor condition. Check for short circuits in load, a broken load, or an open circuit in the load	Shortened battery life possible, further damage to loads.	Repair or replace load, check load, follow up with manufacturer/ system installer for service information

Source NFEC (1989)

- Improper selection of wires (in terms of ampacity, insulation, and so on)
- Unsafe wiring techniques
- Use of inefficient load appliances
- Unsuitable over current protective devices
- Poor system grounding
- Use of more load than what it is designed for
- Difficult to follow O&M (operation and maintenance) instructions

Conclusion

The objective of providing information on maintenance of the systems and general troubleshooting was to give an overview of what all can probably go wrong with a PV system. Most of these problems are simple and can be solved by the users themselves, but it is always advisable that a trained technician looks into these problems.

Reference

NFEC (Naval Facilities Engineering Command). 1989
Maintenance and Operation of Photovoltaic Systems
Virginia: NFEC

Overview of solar photovoltaic programme in India

Historical perspective

Use of solar cells for space applications began in the late fifties. A few national laboratories in India too were engaged in R&D (research and development) on solar cells for a similar purpose till 1975. It was only in 1975 that the DST (Department of Science and Technology) launched a national programme on terrestrial use of SPV (solar photovoltaics). Within the ambit of this, an integrated approach was evolved for the following purposes.

- R&D
- Manufacture
- Deployment and demonstration
- Education and training

To realize this multi-pronged objective, it was decided to actively involve national research laboratories, academic institutions, and the public sector electronics company CEL (Central Electronics Ltd). CEL was entrusted with the task of product development, while others got involved with scientific research on solar cells and material development. From 1975 to 1981, various organizations like the DST, the CSIR (Council of Scientific and Industrial Research), and the DoE (Department of Electronics) offered the much needed support for the Indian SPV programme. This was followed up by setting up CASE (Commission of Additional Sources of Energy) under the administrative control of the DST and later, by a full-fledged DNES (Department of Non-Conventional Energy Sources) in 1982, now MNRE.

CEL began with the processing of 38-mm-diameter hyper-pure silicon wafers using vacuum metallization technique as early as in 1978. Thereafter, the CEL developed an in-house process know-how to develop 100-mm-diameter p-n junction solar cells using low-cost processes like texturization and screen-printed silver metallization. Around the same time, one more public sector company –

BHEL (Bharat Heavy Electricals Ltd) – set up an almost identical technological-cum-production base. The cell efficiencies reached 12%–13% in the production setup with satisfactory yields both at CEL and BHEL. The annual production of solar cells and modules by CEL was 1 MW_p (megawatt peak) during 1990/91. There were just very few manufacturers of cells and modules in the country around that time. Yet another public sector enterprise – REIL (Rajasthan Electronics and Instruments Ltd) – was just producing modules based on technical know-how transferred to it by the CEL.

Crystalline silicon is still regarded as the most suitable material for solar cell fabrication. Thin film amorphous silicon cells are yet to be produced commercially in India. The key to a low-cost single or polycrystalline silicon material is a low-cost but adequately pure polysilicon feedstock. The early PV cells were made from electronic grade polysilcion priced at around \$60–70 in the international market. A major breakthrough was achieved in 1986, when a 25 TPA (tonnes per annum) plant for producing polysilicon was established at Metkem Silicon Ltd. It followed the silicon tetrachloride route based on the technology developed at the IISc (Indian Institute of Sciences), Bangalore.

Metkem silicon also set up facilities for growth of single crystalline silicon ingots and slicing these to produce wafers. Silitronics India Ltd and Super Semiconductors Pvt. Ltd were two more indigenous manufacturers of single crystal silicon wafers in the country. The total single crystal production capacity in 1988 was around 12 TPA. It is pertinent to mention here that nearly earlier, all the requirement of silicon wafers for PV production was import dependent.

During the decade of 1981–91, substantial indigenous capability was built with regard to silicon material, cells, modules, and systems. A variety of PV systems were developed and deployed in the field for the purpose of demonstration and for field-testing and evaluation. More than 40 000 PV systems, ranging from lighting units to village-based power plants, were put up in almost all the states and union territories. In fact, such a demonstration programme came to be known as one of the largest PV deployment programmes worldwide.

The estimated turnover of the Indian PV industry was about Rs 500 million in 1990/91, with about 20 small-scale units engaged mainly in the supply and installation of PV systems. Almost 30% of the total PV activity in the country had attained commercial status during 1990/91. The price of PV modules in India was much more than the price of international modules, mainly on account of higher input costs, duties, and tax structure. The government-administered price for modules deployed under the demonstration programme was fixed at Rs 225 per peak watt and slightly more for the open market programmes.

Recent developments

Lately, the Indian PV industry has led from the front in improving cell efficiencies. Today, commercially produced cells use up to 156 mm × 156 mm sized silicon wafers. Solar to electric conversion efficiencies have now reached 15%–16% owing to significant improvements in the processing technology over the last several years. Silicon wafers continue to be expensive and one of the very effective ways is to use it in a lesser quantity. Present-day wafers are about 200–240 microns thick, as against 350–380 microns thick earlier. Fabrication of silicon wafers is an energy-intensive process too. The direct energy consumption in the manufacture of polysilicon raw material has gradually come down from 330 kWh (kilowatt-hour)/kg in 2003 to around 210 kWh/kg in 2007/08. It has also enhanced the production capacity to about 40 MT (million tonnes) per annum.

Commercialization of thin film solar cells based on amorphous silicon is still waiting to happen. Polycrystalline thin film cells based on cadmium telluride (CdTe) and copper indium diselenide is on the same footing, being still confined to selective few R&D efforts.

PV applications

Lighting

Lighting continues to be a major PV application, having made inroads with the development of a solar street lighting system. A typical system of this type now uses a 74-W (watt) module, together with a 75/80-Ah (ampere-hour) 12-V (volt) lead–acid battery as against a 27-W module used in the early days of the PV programme in the country. The modern day system uses an automatic control to switch ON or OFF a CFL (11 W) instead of a mechanically operated timer before. Likewise, there are now five models of home lighting system, which use PV modules in the capacity range of 18–74 W_p (watt peak), along with a lead–acid battery of 20–80 Ah capacity. There are several models of solar lanterns available in the market, which use PV module capacity of 5–10 W_p, along with a CFL (compact fluorescent lamp) of 5/7 W and a sealed maintenance-free battery of up to 12V, 7 Ah. The LED (light emitting diode)-based solar lantern is a new addition in this category.

Water pumping

After lighting, water pumping is the next widely used PV application. The initial model of a solar pump used a 300–360 W_p array to run a 0.25-HP DC motor pumpset. Subsequently, a higher capacity motor pumpset using a 1-HP DC motor powered by a 900-W array was developed. The total expected discharge was around 77 000 litres from a total head of 10 m and was deemed useful to irrigate about 1.5 hectares of land using surface irrigation and up to 7.2 hectares using drip irrigation technique. It was in 2000/01 that the first successful large project was targeted to deploy around 500 solar pumps using an 1800-W PV array capacity. The purpose was to energize a 2-HP DC motor pumpset. The system was found to be capable of delivering around 140 000 litres of water per day and sufficient enough to irrigate about 5–8 acres of landholding for several crops.

Community facilities

Small power plants, also known as village power packs, continue to be a useful PV system for remote rural communities. These are believed to be competitive with grid extensions in locations separated by about 3–5 km, from the conventional electricity grid. Such power plants in the capacity range of 1–200 kW_p have been installed across the country till now, with West Bengal emerging as a major destination. So far, 23 solar plants totalling 1.2 MW_p (megawatt peak) have been set up in the Sunderbans area of the state to benefit more than 5000 families. Some of the major installations include the following.

- 100-kW_p-capacity system installed in Durbuk block in Leh (Ladakh division of Jammu and Kashmir)
- 200-kW_p-capacity system at Brahm Kumaris Education Society at Manesar in Gurgaon, Haryana

Grid-interactive PV power plants

The MNRE (Ministry of New and Renewable Energy) launched a scheme to augment and supplement grid power by installing PV power projects in the range of 25–100 kW for two niche applications.

1 Rooftop systems on the public buildings to demonstrate peak load shaving applications in major urban centres
2 Distributed grid T&D (transmission and distribution) support systems in remote rural areas at tail-end grid sections

Improvements in tail-end voltage support by 7–15 V were seen in the rural grid. Under this programme, solar-diesel local grid systems in several islands of Lakshadweep were also installed. Seven plants of 750 kW_p capacity have been set up in Lakshadweep islands, besides several such plants of 25 kW_p each standing tall on rooftops of a few government buildings. Within the ambit of this project, a subsidy equal to two-thirds of the project cost, subject to a maximum of Rs 20 million per 100 kW capacity was provided to the state electricity boards, state nodal agencies, and private utilities.

The project was finally closed for a variety of reasons. More recently, the ministry has formulated a major scheme to boost the installation of grid-connected power plants based on a feed-in-tariff mechanism detailed in the following section

Commercial applications

PV for telecommunications is a well-proven market-driven application of high value. India witnessed a major expansion of rural telecommunications network during the early nineties. Under this programme, about 220 000 small PV power systems (generally 70 W_p capacity) were installed to run rural radio telephones. This led to a major expansion of the local PV industry too. However, demand of this magnitude is yet to be seen again.

PV products are seeing an increasing demand for the supply of uninterrupted and unmanned power supply to the cellphone towers. With mobile telephony stretching itself to every nook and corner of India, PV power packs are expected to be much more visible. The IT (information technology) wave is sweeping the country, which also means that PV is set to play a major role by way of providing back-up power supply for computers in rural areas. New areas of PV use may well be the oil and gas, education, housing, IT and eco-tourism sector. Building-integrated use of PV is also expected to increase with a solar housing colony in Kolkata as a model. There has been a steady demand for solar systems for running of lights and television sets, especially in the southern region of the country. Reliable estimates point to a market size of around 8–10 MW_p per year for such applications. The Indian Railways may well turn out to be a major user of PV system for lighting up all its unmanned railway crossings, which are quite accident-prone otherwise.

PV use is no longer restricted to rural areas alone. More and more urban areas are coming within the ambit of solar-based street lighting, signboard displays, and traffic lighting. Quite a few private real estate developers and residents welfare associations are taking pride in installing solar-powered lights in gardens and parks. This sector is expected to grow further with the municipal corporations in various states like Delhi, Haryana, Karnataka, and Maharashtra going in for these types of systems. As environmental consciousness is increasing, efforts are being made to replace the noisy and smoky small diesel-/petrol-operated gensets (up to 10 kVa [kilovolt-ampere]) by PV power packs in urban centres.

PV market development programme in India

The MNRE is the apex organization in India for policy-making, planning, promotion, and coordination of the different aspects of renewable energy. In the initial stages, it put in place various field-level programmes to showcase the usefulness of PV technology and thereby a number of products and systems. Further impetus was given to the programme by setting up of state nodal agencies for renewable energy, along with direct marketing solar shops in various states across the country. Despite running one of the largest PV demonstration programmes in the world, it was still not an easy task to sell the PV systems. Taking an early note of this vexing problem, the MNRE set up IREDA (Indian Renewable Energy Development Agency) in 1987, which was delegated with the responsibility to commercialize PV technology in 1993/94.

Key PV market segments

There are several PV products and systems available in the market today. Lighting products constitute the bulk of the need-based products in sizeable market demand. In essence, there are four potential types of markets for PV systems. They are (1) Government, (2) Government-driven, (3) Private leasing, and (4) Direct sales (on the open market)

PV products have gained increased field performance reliability, a definite improvement over the products delivered under the early phase of the PV demonstration programme. The resultant effect was that private markets began to take shape, which needed reasonably worked out incentive structures. PV manufacturing capability, too, was reinforced through a series of measures undertaken by IREDA. Nearly 20 PV units producing various systems were offered financial support. As of now, the PV industry has matured with time and experience. Today, the cumulative PV capacity in the country has touched 335 MW, out of which exports account for as much as 225 MW.

PV marketing models

There are different ways in which PV systems are sold in the market. IREDA (Indian Renewable Energy Development Agency) makes use of intermediaries to offer loans to the consumers within the ambit of the following four finance models.

1   The cooperative model uses rental and leasing mechanisms.
2   The corporate model uses hire-purchase and leasing mechanisms.
3   The NGO model uses a combination of rental, hire-purchase, and leasing mechanisms.
4   The dealer model uses direct sales to the user.

A significant objective of the above models is to enable cross-section of users to get PV systems from the local suppliers at an affordable price paid back in easy instalments.

Major PV projects

PV water pumping

Water pumping is a widely used PV application. IREDA implemented a national PV water-pumping programme on behalf of MNRE with a clear objective to demonstrate the use of PV pumping systems in agriculture and related uses. Other sectors of use included horticulture, animal husbandry, poultry farming, high-value crop production, orchards, silviculture and importantly, drinking water supplies. Till now, more than 7068 water-pumping systems catering to the above applications have been installed across the country so far. Under this specific programme, financing companies, and the system manufacturers/suppliers function as intermediaries between IREDA and users by offering leverage on the high initial costs of products.

SPV market development

The World Bank-supported-project on PV market development in India came at an opportune time when sufficient number of PV products had already been demonstrated in the field under various government programmes. To begin with, a countrywide market survey of the potential PV products ranging from very low-power consumer products to hybrid power systems was commissioned. IREDA was the implementing organization for this prestigious market development project. In all, 85 projects with an aggregated solar power generating capacity of 2.1 MW were supported. Table 1 shows the types of systems delivered across various geographical regions in the country under such a programme.

Table 1 Types of PV systems delivered across various geographical regions

Product/system	Number
Solar lantern	39 000
Home lighting system	3 602
Street lighting system	1 016
Garden lighting systems	349
Solar water-pumping systems	86
Stand-alone power plants	8

Source IREDA (Indian Renewable Energy Development Agency)

Special concessions were given to motivate entrepreneurs to establish PV projects in remote rural areas. These mainly included the following.
- Provision for higher outlay from IREDA (means less of contributions from entrepreneurs)
- Reduced interest rates
- Waiving of commitment fees and inspection charges

Capacity building and training (under World Bank project)

PV performance specifications in respect of marketworthy products and systems were prepared by IREDA for adherence by the corporate entities as also the NGOs. Besides, requisite awareness generation was created amongst the potential customer segments through all possible ways, including business meetings, workshops, seminars, exhibitions and audio-visual films. IREDA made special arrangements to develop a workforce of about 900 trained engineers and technicians. Siemens Solar was chosen as the agency for imparting the desired practical skills and the whole programme was organized at the CEC (Central Electronics Centre), IIT Madras.

Photovoltaic market transformation initiative

SHS (solar home systems) are fast gaining market acceptance amongst individual households, small hospitals, community centres, schools, and many more segments. This is primarily due to a large number of rural areas being without power, in addition to those regions with quite low availability of daily power supply. Such systems are also being used to run low-power appliances like radio and TV, which are much in demand. Not only that, SHS increase the working hours into the night for an income-generation activity like basket weaving. However, it has not been easy to market these solar home systems. The initial perception of such systems performing poorly, together with the near total absence of a reliable after-sales network and easy loan financing affected the market sales to a large extent. However, the scenario began to change for the better with the launch of a major PV marketing project under the now well-known acronym PVMTI (Photovoltaic Market Transformation Initiative).

PVMTI is funded by the IFC (International Finance Corporation), along with GEF (Global Environment Facility), with a clear-cut mandate to promote sustainable commercialization of PV technology. The developing world has been chosen for introducing of successful and replicable business models. Between 1998 and 2007, an amount of around $1.6 million has been committed for more than nine projects in India, Kenya, and Morocco. In India, five projects were financed with a commitment of $15 million in the form of debt, grants, equity, and guarantees. Table 2 shows a summary of PVMTI-funded projects in India.

Table 2 PVMTI-funded projects in India

Company (invested in)	PVMTI investment ($ million)	Project brief
Shri Shakti Alternative Energy	2.2	Expansion of a network of energy stores in southern region of India, marketing of PV, along with other renewable energy products
SREI International Finance	3.5	Introduction of an innovative credit scheme for PV customers in West Bengal
Shell Solar India	4.0	Expansion of a network of Shell Solar centres in South India, in addition to establishment of a credit scheme for end-users in partnership with local financial institutions
Aqua Solar	3.0	Setting up of a network of PV-powered water-pumping, purification, and bottling stations in rural India
Sunlit Media Solutions	2.3	Setting up of a network of PV-powered advertising signs

PVMTI – Photovoltaic Market Transformation Initiative

Source www.pvmti.com

UNEP Solar Loan Project in India

This four-year project worth $7.6 million was launched in April 2003 to stimulate the low sales volume of solar home systems. UNEP (United Nations Environment Programme), together with the Shell Foundation, supported this major innovative financing project. It turned out to be a rewarding partnership involving UNEP, the UNEP Risoe Centre, and two of India's major banking groups—Canara Bank and Syndicate Bank, along with their sponsored grameen banks. The approach adopted by UNEP was centred on finding banking partners who were keen to develop this new sector with some support. The project was executed in South India in locations that had little or no electricity. The following key project planning steps were involved.

- Reduced interest rates
- Market development support
- Process to qualify the solar vendors

The interest rate reduction was phased out during the programme and the market of solar home systems was run on purely commercial terms. Highlights of this global award winning project are given in Table 3.

Table 3 UNEP Solar Loan Project

Number of households involved	18 000
Approximate number of people benefited	100 000
Total number of loans disbursed	19 533
Total number of bank branches (participated)	2076
Total number of solar vendors qualified	05*

*Kotak Urja Private Limited, Omega Electronics, Selco Solar Electric Light Fund, Shell Solar India Private Limited, Tata–BP Solar India Limited

Source www.unep.org

Successfully adopted PV marketing mechanisms

PV systems are expensive, but that does not mean using the medium of subsidy alone to sell such systems. A number of innovative marketing approaches have been devised for the purpose. The following are regarded as the most suitable techniques to take PV systems to those who cannot afford them.

Setting up credit sales

In this case, loans were offered for buying the solar home systems. Most of such sales are now credit based, and the customers are at liberty to pay for these systems over 3–5 years. The idea is to keep the monthly loan instalment below the amount spent by these poor category people on traditional methods of lighting like a kerosene oil lamp.

Quick and reliable after-sales services

Strong dealer networks have been created, which focus on timely after-sales service. Technicians, who are now specially trained for the purpose, undertake regular visits to the customers premises as per a well-coordinated activity schedule. In quite a few cases, a trained member of the user community does the routine maintenance of the systems.

Working in tandem with the grassroots-level organizations

There is a sizeable number of NGOs (non-governmental organizations), rural cooperatives, and self-help groups involved in community development at the grass-roots level. Those supplying the systems in the manner described above often seek their support and cooperation for the following purposes.

- System installation
- Marketing
- Awareness generation
- Interfacing with potential customers
- Possible financing of the systems (in some cases)

Customizing the systems

Solar systems are typically not off-the-shelf products. Each user can have a requirement that is at variance from the rest. The technicians are now fully equipped to study the site specifics and offer a suitable system choice. For example, one light may be installed in a manner that it lights up part of an adjoining room too.

Efforts at income generation

It is quite logical to find ways and means to create some extra income via the use of solar home systems. The income thus generated can then be used to pay back the loan instalment. A good example would be the purchase of a solar system by a roadside vendor. He/she then enjoys extended hours of product sales under the cool white light of a solar lamp.

Approach for enhanced market penetration

There is a wide cross-section of views on how to enhance the market penetration of solar PV systems in the country. While many would like economies of scale to come into the picture, yet others express the following views.

- Enhance R&D spending to achieve efficiency improvements not only in cells and modules but total system too.
- Catalyse the growth of ESCs (energy servicing companies), which can reach out on different fronts to rural and urban areas alike.
- Encourage grameen banks to function as intermediaries for micro-financing, that is, facilitating micro-financing for the purchase of PV products and systems.
- Carve out well-defined roles for cooperatives and NGOs so as to promote PV in rural and semi-rural areas.
- Increase the participation of corporate bodies and have them act as intermediaries in the leasing and hire purchase of PV systems for a large number of consumers.
- Put in place concessional loan schemes so as to expand the potential customer base.
- Work out favourable fiscal incentives to bring down the long-term costs of PV products.
- Ensure sustained, intense, and regular interaction amongst all stakeholders through all possible ways and means.

PV market trends

The PV market has grown from just a few megawatts in 1990/91 to more than 335 MW_p (megawatt peak) now. Incidentally, exports from India to other countries account for almost 67% of the total market, leaving a local market share of around 33%. It indicates a huge untapped potential for the PV market in the country for various applications. As per reliable estimates, more than 1.4 million PV systems (109 MW_p) have been deployed in the country so far, mainly for small decentralized applications. Table 4 shows the sectoral use of PV cells and modules as in December 2007.

Table 4 Use of PV cells and modules by sector (December 2007)

Product application	Market share (MW)	Market share (%)
Solar lantern	7.5	6.88
Home light	16.5	15.1
Street light	5.5	5.04
Water pumping	11.0	10.09
Power plants	6.0	5.5
Telecom	22.0	20.18
Others	40 .0	36.69

MW – megawatt
Source MNRE (2007)

Quite clearly, lighting applications constitute roughly one-third of the total PV market.

Outlook for the Indian PV industry

From just one MW of PV cell and module in 1990/91, today there are nine companies producing cells and twice as many producing solar modules. PV-specific investment is catching on fast with both the national and international companies. Quite a few reputed foreign PV companies are busy exploring possibilities to set up their production/marketing bases in India. Over the last 15 years or so, several private companies have either scaled up their production capacities or set up new manufacturing initiatives based on cells and modules especially. As of now, the entire production capacity is based on crystalline silicon. The following are the most important indicators of a growing Indian PV industry.

- Liberalized industrial policy for setting up PV manufacturing facilities put in place in 1992.
- Existing installed capacity for solar cells is 155 MW_p with an equivalent capacity present for solar modules.
- 40 MWp of cell production capacity and 100 MW_p of module capacity likely to be added by March 2008.
- Thin film manufacturing initiatives are still limited to assembly of modules.
- Just two companies are producing silicon wafers and ingots with an installed capacity of 4.5 MW_p equivalent.
- Around 50 companies (mostly small scale units) are involved in the manufacture of PV systems.
- In 2006/07, nearly 45 MW_p of solar cells and 80 MW_p of solar modules were actually produced in the country, and the

cumulative production of cells and modules by 31 March 2007 was estimated to be 242 MW_p and 355 MW_p respectively.

- The average annual growth of domestic production of modules is between 20%–25% for the last few years.
- More than 60 MW_p capacity of cells and modules were exported by the Indian manufacturers to serve increased demand from Germany in view of its favourable feed-in-tariff policy.
- Reliance Industries is looking at setting up an over $5-billion worth PV cell manufacturing facility in Jamnagar (Gujarat) besides a few more companies like Signet Solar and MSPL.

Case studies

PV market delivery

In the last few years, there have been several attempts to promote PV systems on full-scale commercial terms. Studies have indicated that the most crucial aspects for promoting these systems in the rural areas are infrastructure for repair and maintenance, and credit facilities. However, efforts to establish such infrastructure, so far have had limited success.

A pioneering effort has been made by SELCO (Solar Electric Light Company) in South India. The company has established a service network by training local entrepreneurs in installation, repair, and maintenance of PV systems. It has also created infrastructure for credit to potential consumers. Yet another approach has been in the Sundarbans area of West Bengal, where the Ramakrishna Mission has initiated the promotion of PV domestic lighting systems under the MNES programme. It has also created an infrastructure for repair and maintenance. Similar attempts have also been made by the NGO Social Work Research Centre, which evolved the concept of 'barefoot engineers', and trained local youth to provide back-up facilities.

SELCO: taking PV to the villages

SELCO India is a solar energy services company based in Bangalore, which has established a process of distribution of PV systems by commercial channels as a profitable business. The main objectives of the company are (1) to provide electricity to the rural population and (2) to integrate PV into the mainstream of the local culture. SELCO essentially assembles and markets PV systems to rural households. It started its ground-level activities in the villages of Karnataka in April 1995, with a small amount of seed capital from the Rockefeller Foundation channelled through a US-based non-profit organization called SELF (Solar Electric Light Fund) Inc.

SELCO has formed a network of 25 energy service centres, and is currently operating in three states—Karnataka, Kerala, and Andhra Pradesh. It has a trained workforce of local technicians and collection agents in the selected villages. The technicians work for SELCO on an income-cum-commission basis with an incentive to sell more systems. Technicians double as salesmen also. The commission depends on whether it is a direct sale or a sale on a loan basis. The aim is to provide quick after-sales services to customers. It also facilitates monthly collection of loans.

The company has a strong focus on market delivery of standalone systems and service in rural areas mainly for lighting. This strategy of SELCO has helped the company to establish beneficial relationships with nationalized banks, grameen banks, NGOs, IREDA, and international investing agencies for the promotion of PV systems.

SELCO's marketing models

SELCO has created different marketing mechanisms to suit the various categories of its customers. These mechanisms can be classified into four types.

1. Cash sale (direct sale to customers)
2. Sale through nationalized/grameen bank (loan to customers)
3. Sale through IREDA (system on lease-to-own basis)
4. Sale through local institutions (marketing of systems through local institutions, namely, cooperatives, farmer societies, and NGOs)

Cumulative achievements of SELCO

The company has gained an impressive foothold in the Indian PV market scene and was recently conferred the prestigious ASHDEN award for reaching out to the poor sections of the society. The major achievements of SELCO India so far may be summarized as below.

- Installation of more than 75 000 systems in rural areas
- Establishment of an extensive network of 25 energy service centres
- Provision of the basic benefit of lighting to more than 300 000 people
- Profitability on annual sales in excess of $3 million
- Creation of a fully trained staff of about 200 people at various skill levels

NGO approach: Ramakrishna Mission

The strategy adopted by the Ramakrishna Mission for the promotion of PV technology in rural areas is another unique experience. The Lok Shiksha Parishad of the Mission has a separate unit – the SEU (Solar Energy Unit) – which works on solar energy. The SEU looks after the installation, repair, and maintenance of PV systems. It also manages the ADITYA shop, which is supported by the MNRE and WBREDA (West Bengal Renewable Energy Development Agency). It has so far installed more than 1332 home lighting systems in the Sundarbans, Midnapore, and parts of 24 Parganas(s) district.

Infrastructure for PV promotion

At the SEU, there is a team of technicians and trained staff that is responsible for the installation of the systems at different sites. The team also orients the visitors to the solar shop on the applications of the different systems and how these can be purchased. The Ramakrishna Mission implements the projects through the affiliate block-level cluster organizations. However, only a few of these are involved in the installation of the PV systems. Each of these organizations has a team of skilled technicians who assist the SEU team in installing the systems and also provide technical support to the village youth clubs. The village youth clubs are involved in the implementation of the projects.

Maintenance and repairs

The SEU looks after the major repair and maintenance of the systems. In most village youth clubs, trained technicians have been appointed to undertake minor repairs and maintenance. If, the faults in the systems cannot be repaired by the village technician, these are directed to block-level organizations. Most of the faults in the systems can be repaired in block-level organizations. If the system cannot be repaired at the block level, it is taken to the SEU.

Dissemination

Village youth clubs, block-level organizations, and the SEU are the dissemination centres. Potential buyers approach these centres and apply for the systems. In case the buyer needs credit, the block-level organization assists him/her in getting a loan from the commercial/local area banks. Village youth clubs and block-level organizations also conduct awareness programmes for the villagers and demonstrate the systems during *kisan/krishi melas* (agricultural fairs).

Reasons for success

The SPV programme of the Ramakrishna Mission has achieved enormous success. It promotes the use of not only solar PV systems, but also biogas plants. This has been possible primarily because of the following two reasons.

1 Strong infrastructure for dissemination, repair, and maintenance
2 Credibility of the organization

Most recent PV policy initiatives

Incentive scheme for solar power projects

The MNRE has just formulated a well-focused scheme to encourage the use of PV for large-scale power generation. Key objectives of this scheme are as under.

- It is a demonstration programme to support megawatt capacity grid-interactive solar PV power generation projects up to a maximum capacity of 50 MW in the country.
- Any state will have a maximum of 10 MW of solar PV power capacity under this scheme.
- Projects in a minimum size of 250 kW_p will be considered, which can later be scaled up to 1 MW.
- In respect of projects approved and commissioned by 31 December 2009, the ministry will offer generation-based incentive of Rs 12 per unit for solar PV projects. This will be done after taking into account the tariff provided by SERC (state electricity regulatory commission) or the state utility.

Special incentive package to encourage investment in semiconductor fabrication

The Ministry of Communications and Information Technology has come up with a well intended package of benefits to boost semiconductor production in the country. Significant objectives of this scheme are as follows.

- The investment is for the production of solar cells and devices too.
- The threshold NPV (net present value) of the investment for manufacture of eco-systems (such as liquid crystal displays, organic LED, and solar PV) has been fixed at Rs 1000 crore (10 billion) and above.

The central government or any of its agencies will provide incentive of 20% of the total capital expenditure during the first

10 years for the units falling in the SEZs (special economic zones) and 25% of the capital expenditure for the non-SEZ units. Non-SEZ units will be exempted from countervailing duty.

- The capital expenditure will be the total of expenditure on land (2% of the total), building, plant, and machinery and technology.

Development of solar cities

The MNRE is now making all out efforts to popularize the use of solar energy technologies in the urban areas. A total outlay of Rs 30 crore (300 million) has been earmarked under this novel scheme. The major objectives are to

- empower urban local bodies to address challenges at the city level;
- create solar capacity in an urban ambience;
- develop a participatory approach of programme implementation amongst all those concerned; and
- set up a total of 60 cities as solar cities.

Conclusion

Several attempts have been made so far to increase the market share of PV systems. Quite a few of the market promotion approaches have been successful in their respective geographical locations. The key aspect to the market promotion of PV systems today is to create infrastructure for financing and effective after-sales service. The approaches adopted by the Ramakrishna Mission and SELCO have been successful largely because they have addressed the problem of repair and maintenance. Another key aspect to the promotion of PV systems in rural areas is the availability of credit. If these two aspects of market development are taken care of, there is a vast market of PV systems. It is in this context that programmes like PVMTI and the UNEP solar initiative assume significance.

PV systems are already considered as one of the best options for rural infrastructure projects in unelectrified areas. It is expected that the cost of PV modules will come down significantly, and this will happen with further expansion of the market. Solar energy is not very competitive now, but higher volumes and improved technologies are expected to bring down costs considerably.

Reference

MNRE (Ministry of New and Renewable Energy). 2007
25 Years of Renewable Energy in India
New Delhi: MNRE, Government of India

Overview of the international photovoltaic programme

Introduction

Solar PV (photovoltaic) technology is making its presence felt both in the developed and developing regions of the world for a variety of reasons, though the demand for PV electricity is the highest in developing countries. In the year 2000, the cumulative installed capacity of the solar PV systems worldwide was around 1200 MW_P (megawatt peak). Since then it has increased more than five-fold to touch 6500 MW_P. The average annual rate of growth of solar PV has been more than 35% for the last two decades or so.

Today, the global solar industry has an annual market worth in excess of \$5.8 billion. New production facilities and technologies are being set up throughout the world, more so in Japan and Europe. India too is witnessing a growing interest in the PV specific investment in the form of individual and joint ventures. Of special mention is the fast-track growth of PV programmes in countries like Germany, Spain, and the USA, all due to favourable policy frameworks.

Silicon feedstock

Crystalline silicon continues to be the workhorse of the global PV industry. The demand for silicon wafers is at an all-time high in view of the expanding frontiers of the PV market. Till now, nearly 50% of the global production of the electronic grade silicon was directly used by the PV industry. However, new production capacities based on slightly lower purity grade of electronic silicon are now taking shape. As per reliable estimates, the projected outlay for upscaling PV silicon production capacities during 2007–10 is in the range of approximately \$2.5 billion or even higher. The total anticipated worldwide market demand in 2008 is expected to be around 3800 MW_P.

Brief analysis of key global PV markets

The following section gives a quick insight into the growth patterns of the PV markets in Germany, USA, Japan, Spain, Italy, and China.

Germany

Today, Germany is synonymous with the large-scale demonstration of rooftop PV systems meeting various end-use applications. It is still regarded as the largest market for solar panels or modules, having obtained a global market share of just under 50% in 2007. The key driver responsible for the market leadership position is an incentive model commonly known as the 'feed-in-tariff' system. It simply means that the energy company offers a cost-covering fees in respect of every unit of solar-PV-derived electricity. The costs are taken care of by levying a small surcharge on the units that all German users purchase from the energy company. A positive fallout of this model has catalysed a continuous and imposing growth both within the German PV market and the global PV industry at large. As of now, more than 1500 companies in Germany are associated with this PV programme, ranging from big industrial producers to companies specially trained in system installation. Following a detailed programme review, there has been a reduction of about 5% in the feed-in tariff level during the last few years. It is expected to be lowered down further by 8%–9% in 2009 and thereafter. Market estimates point towards the size of German PV market being little below 2000 MW in 2010.

USA

The US programme enjoyed a distinct market leadership position for many years in the past. Of late, this is no longer the case despite the aggressively pursued subsidy-driven programmes in New Jersey and California. Solar PV programmes in the other states have been initiated through obligations imposed upon energy companies. It means that such companies need to sustainably produce or purchase a percentage of the energy sold. Few states have clearly earmarked that percentage for solar energy. As per available market estimates, the PV market in California recorded an approximate growth of 50% in 2007. The federal tax credit is due for expiry by the end of 2008, and nothing seems certain about its continuation in 2009. However, as per very optimistic estimates of the growth rate, the US market may approach around 1500 MW_p in 2010.

Japan

The PV rooftop programme was an exciting programme watched with keen interest worldwide. Of late, its market share has slid, with hardly any growth in 2006 while dipping further in 2007. At present, it represents about 8% of the global PV market. It seems the termination of a national subsidy programme in 2005 led to subdued market growth in the following years. Subsequent to this, the Japanese industry pursued a keen interest in exporting its PV range of products. There are some signs of new incentive schemes being introduced, pending which a moderate growth rate of 20%–30% may be witnessed in this unsubsidized market.

Spain

Perhaps the most spectacular growth of the PV market in recent times has been recorded in Spain. It experienced an approximate growth of 300%, to attain around 426 MW_p of the new installed PV capacity. The main reason for this jump is the availability of an attractive feed-in-tariff law, which encouraged many overseas investors to pump their money into Spain. In the process, many new solar parks of a few megawatts were constructed. The above-mentioned scheme is due for closure in September 2008, which anyway has led to a strong market growth by now. The Spanish market is expected to account for as much as 22% of the global PV market in 2008.

Italy

Italy is a fast emerging PV market in Europe, signalling a positive market growth by 2012. Here too, an attractive feed-in-tariff scheme is in place, which has resulted in new capacity addition of around 70 MW in 2007. Market estimates indicate an intense market growth of almost 100% or even more in 2008. Obviously, a large number of international investors are rushing to Italy to push up the market further.

China

The Chinese market is still developing, with a majority of solar cells being used for rural electrification (41.3%) followed closely by communications, and industrial applications (33.8%). The total installed capacity in 2006 was around 80 MW_p, out of which nearly 53.8% finds use in the commercial sector (telecommunication, industrial applications, and solar appliances like road signs). The remaining capacity – 46.2% – sustains solely due to government support (rural electrification and grid-connected generation). On the solar cell production front, the European market, especially in Germany, boosted the development of production capacity in China. Most of the modules produced in China found their way into Germany, with just about 5–10 MW_p having been installed in China in 2005/06.

It is quite clear that a stable and secure scheme is sufficient to ensure a market growth of more than 100%, as witnessed in case of Germany and Spain.

Production and distribution points for solar cells and modules

Silicon wafer is the basic material for producing a solar cell and therefrom a solar module. The USA, Europe, and Japan have traditionally figured amongst the bulk producers of solar cells and modules. However, with favourable policy formulations and incentives, the overall market dynamics has begun to change, more so over the last few years. Table 1 gives an idea about the major global cell producing regions, while Table 2 lists the top solar cell producing companies of the world.

Table 1 Global cell production (MW_p)

Country	2000	2001	2002	2003	2004	2005	2006
US	75.0	100.3	120.6	103.0	138.7	154.0	201.6
Japan	128.6	171.2	251.1	363.9	601.5	833.0	926.9
Europe	49.8	73.9	122.1	200.2	311.8	476.6	678.3
Rest of world	23.4	40.6	53.3	81.3	141.5	322.5	714.0
Total	276.8	386.0	547.0	784.4	1193.5	1782.3	2520.7

Source PV News, April 2007

Japan occupied the highest share of cell production in 2006, followed closely by Europe.

The next logical step is the production of modules, which are then used for meeting various end-use applications. Table 3 gives the geographical distribution of modules when being used in the field.

Table 2 Market share of top 10 PV cell-producing companies

Company	Country of origin	Cell production (MW) (2006)	Cell production (MW) (January–June 2007)
Sharp	Japan	434	225
Q-Cells	Germany	253	417*
Suntech	China	158	145
Kyocera	Japan	180	108
Motech	Taiwan	102	85
Sanyo	Japan	155	87
Deutsche Solar	US/Germany	86	66
First Solar	US	60	61
Mitsubishi	Japan	111	55
Sun Power	The Philippines	63	54

MW – megawatt
* As on 31 December 2007
Source Earth Policy Institute Report 2008

Table 3 Global PV module installations (systems only) in MW_p

Country	2000	2001	2002	2003	2004	2005	2006	2007*
Germany	44	78	80	170	500	700	1050	1260
Rest of Europe	1	3	10	11	24	60	118	234
Japan	74.4	91	141	201	256	320	350	402.5
Rest of Asia	13	19	43	33	47	55	81	131.9
US	16.8	28.4	49.1	71.7	89.9	108	141.4	259
Rest of world	21	37	77	118	147	139	130	158.8
Total	170.2	256.4	400.1	604.7	1063.9	1382	1870.4	2446.2

MW_p – megawatt peak
* estimated
Source PV News, July 2007

PV applications

As of now, about 1.6 billion people around the world do not have
access to basic electricity, nearly 80% of them in rural areas. Solar
lighting systems in particular can ensure a better quality of life for
such deprived people. Lighting, water pumping and battery charging
for running TV, fans, and radio are amongst the most widespread
uses of PV technology. PV vaccine refrigeration systems are equally
in demand. Reliable estimates point out that around 7%–8% of the
global PV installations accounted for rural electrification purposes
in 2006. Developing countries are witnessing large-scale deployment
of solar home systems. By 2007, more than 2.5 million households
in countries like Nepal, China, Bangladesh, India, Sri Lanka, and

Thailand were benefiting from the use of home lighting systems. Market delivery mechanisms for such systems have received support from micro-credit and donor programmes.

As far as urban uses of PV are concerned, it is visible in the form of running very low-power consumer products and small devices like calculators, ventilators, road signs, water sprinklers, and power for phone boxes. As per reliable estimates, these products made up nearly 25% of the PV demand during 2006/07. This sector may continue to grow in a near identical manner considering the growing number of such devices in use. Table 4 gives a break-up by application of the cell/module use, ranging from milliwatts to megawatts.

Table 4 Global PV cell use by application

Market sector	1996	1998	2000	2002	2004	2005	2006
Consumer products	22	30	40	60	75	80	90
World off-grid rural	23	34	53	85	110	125	140
Communications/signal	23	31	40	60	80	90	100
Off-grid commercial	12	20	30	45	55	60	70
Grid-connected residential/commercial	7	35	120	270	700	1375	1600
Large (>500 kW)	2	2	5	5	20	30	100
Actual (MW/year)	89	152	288	525	1040	1550	2200
Actual average module price ($/W$_p$)	4	4	3.5	3.25	3.25	3.5	3.75

kW – kilowatt; MW – megawatt; W$_p$ – watt peak
Source PV News, July 2007

There is now a growing market for the use of PV-grid-connected systems. Design features allow PV modules to be put on the rooftop or architecturally integrated in the roofs and facades of houses, offices, and even public buildings. Use of PV is also being made for replacing the elements in a building envelope. Solar roof tiles or slates can easily substitute the conventional materials. There are more than 800 PV power plants worldwide with capacity larger than 200 kW and at least nine plants with a capacity higher than 10 MW in Germany, Portugal, Spain, and the USA.

Thin film modules offer an additional advantage in the form of convenient design geometry in a building envelope. PV power is also proving useful for bringing down the need for peak power during summer.

The grid-connected PV continues to be the fastest growing power generation technology in the world, with 50% annual increase in cumulative installed capacity both during 2006 and 2007. Estimates point to about 1.5 million homes having PV-rooftop systems worldwide. These feed the solar PV power to the grid, with Germany emerging as a clear winner in this regard. Three other major markets for PV-grid systems are Japan (300 MW), USA (100 MW), and Spain (100 MW_p) in 2006.

It is pertinent to note here that the PV-grid-connected capacity has jumped to 85% in 2006 as against just 20% in 1994. This has largely become possible due to the firm footing of support programmes in countries like the USA, Germany, and Japan. The concept of net metering has fast gained ground and will stay for many more years to come. India, too, is expecting substantial additions to its PV capacity as a result of just notified feed-in-tariff of Rs 12 per unit of PV electricity generated.

Ensuring connectivity between the rural areas is high on the agenda of developing countries. In fact, use of PV for telecommunications is contributing to a very high growth of the PV industry. More recently, repeater stations for mobile telephony in countries like India are being run on solar power. Other important off-grid uses of PV are in meteorological stations, traffic lighting, and water desalination. Finally, it is of definite interest to take a close look at how the global PV capacity has added up all these years say during the last 15 years or so. Table 5 shows such cumulative PV capacity.

Table 5 Cumulative PV capacity

Year	Capacity (MW_p)	Growth during successive years (%)
1994	502	
1995	580	15.5
1996	669	15.3
1997	795	18.8
1998	948	19.2
1999	1150	21.3
2000	1428	24.1
2001	1762	23.3
2002	2201	24.9
2003	2795	26.9
2004	3847	37.6
2005	5167	34.3
2006	6634	28.3
2007	9740	46.8

MW_p – megawatt peak
Source Greenpeace (2007), www.worldwatch.org

Overview of the international photovoltaic programme

Global PV market forecast for 2008

Germany is quite likely to retain its market leadership position in 2008, followed by Spain. The erstwhile market leaders like the US and Japan lag far behind Germany. Figure 1 showcases the major PV markets worldwide.

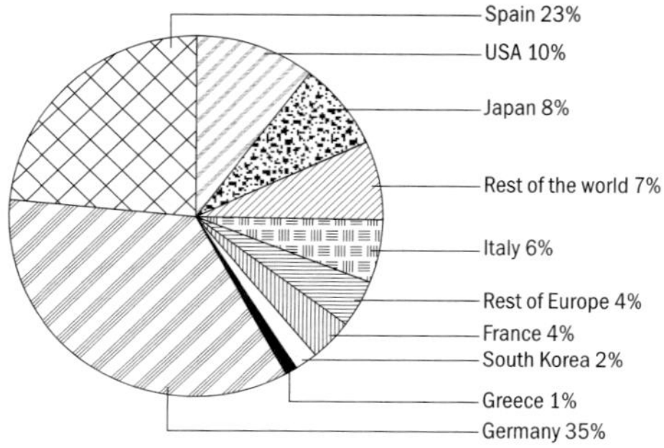

Figure 1 Global PV market projections in 2008

Conclusion

The global PV market has traditionally been regarded as a stronghold of government-instituted subsidy programmes. Seemingly, PV technology will gain acceptance only when it gains parity with grid power. PV enthusiasts are eagerly waiting for that level of development, keeping in view the technological innovations currently under way in several places. The PV industry on its part is trying to attain economies of scale through major capacity expansions to realize the ultimate objective of lower cost per peak watt. Likewise, the multi-pronged initiatives triggered by an altering climate change scenario may finally tilt the balance in favour of PV.

Reference

Greenpeace. 2007
Solar Generation IV-2007
Details available at http://www.epia.org/fileadmin/EPIA_docs/publications/epia/EPIA_SG_IV_final.pdf, last accessed on 5 September 2008

Bibliography

CEA (Central Electricity Authority). Various years
Annual Report
New Delhi: CEA

Derrick A, Barlow R W, McNelis B, Gregory J A. 1993
Photovoltaics: a market overview
London: James and James Science Publishers Ltd

Gregory J, Silveira S, Derrick A, Cowley P, Allinson C, Paish O. 1997
Financing Renewable Energy Projects: a guide for development workers
London: IT (Intermediate Technology) Publications Ltd

Hande H. 1995.
SELCO Model for Solar Rural Electrification
Mangalore: Solar Electric Lighting Company

Harald N R. 1992
The Sunshine Revolution
Stavanger: Sun Lab Publishers

IREDA (Indian Renewable Energy Development Agency). Various years
Annual Report
New Delhi: IREDA

IT (Intermediate Technology) Publications Ltd. 1997
Electricity from Sunlight: photovoltaic applications for developing countries
London: IT Publications Ltd

Kulsum A. 1994
Renewable Energy Technologies: a review of the status and costs of selected technologies
Washington DC: World Bank

Laws R J. 1983
Solar Cells: what you always wanted to know
New Jersey: Enslow Publishers

Lorenzo E. 1997
Photovoltaic rural electrification
Progress in Photovoltaics: research and applications 5: 3–27

Louineau J P, Modibo D, Fraenkel P, Barlow R, Bokalders V. 1994
Rural Lighting: a guide for development workers
London: Intermediate Technology Publications Ltd

Ministry of Power. Various years
Annual Report
New Delhi: Ministry of Power, Government of India

Merrigan J A. 1976
Sunlight to Electricity: prospects for solar energy conversion by photovoltaics, 2nd edn
Massachusetts: MIT Press

MNRE (Ministry of New and Renewable Energy). Various years
Annual Report
New Delhi: MNRE, Government of India

NFEC (Naval Facilities Engineering Command). 1989
Maintenance and Operation of Photovoltaic Systems
Virginia: NFEC

Northrop M F, Riggs P W, Raymond F A. 1995
Setting Solar: financing solar energy in the developing world
[Pocantico Paper No. 2, the Pocantico. Conference Center of the
Rockefeller Brothers Fund]

Parthasarathi A, Madhavan D, Kaul V K. 1994
Solar photovoltaic energy products: from semiconductor physics to advanced energy technology
[Paper presented at the International Conference on Physics
and Industrial Development: bridging the gap, New Delhi, 17–19
January 1994]

Paul L J, Dicko M, Fraenkel P, Barlow R, Bokalders V. 1994
Rural Lighting: a guide for development workers
London: Intermediate Technology Publications Ltd

REC (Rural Electrification Corporation). Various years
Annual Report
New Delhi: REC

Rostvik H N. 1992
The Sunshine Revolution
Norway: Sun Lab Publishers

Sinha S. 1997
**Financing of renewable energy technologies: financial
organization perspective**
TERI Newswire **III** (23): 5–7

TERI (Tata Energy Research Institute). 1997
PV credit mobilization
New Delhi: TERI
[Report No. 96/RE/56]

Turner R P. 1975
Solar Cells and Photocells
Indianapolis: Howard W Sams & Company

UNESCO (United Nations Educational, Scientific, and Cultural
Organization). 1988
Photovoltaics for Rural Electrification
Paris: UNESCO

Wim H and Peter F. 1994
*The Power Guide: an international catalogue of small-scale
energy equipment,* 2nd edn
London: Intermediate Technology Publications Ltd

Websites

Akshay Urja
http://www.mnre.gov.in/akshayurja/contents.htm

China Solar PV Report 2007
http://www.greenpeace.org/china/en/press/reports/china-pv-
report

IREDA News
http://indianjournals.com/ijor.aspx?target=ijor:in&type=home

REEEP (Renewable Energy and Energy Efficiency Partnership)
http://www.reeep.org/

Renewables 2007 Global Status Report
http://www.ren21.net/pdf/RE2007_Global_Status_Report.pdf

Solar Energy Society of India
http://www.ises.org/india/

The Economic Times
http://economictimes.indiatimes.com/

The Financial Express
http://www.financialexpress.com/

The Hindu Business Line
http://www.thehindubusinessline.com/

The Mint
http://www.livemint.com/Home.aspx

Glossary

Air mass An indication of the length of the path solar radiation travels through the atmosphere (An air mass of 1.0 means the sun is directly overhead and the radiation travels through one atmosphere [thickness]).

Alternating current (AC) Electric current that reverses direction at frequent intervals, usually many times a second.

Ambient temperature Temperature of the surrounding area.

Ampere-hour A measure of the flow of current (in amperes) over one hour; used to measure battery capacity.

Ampere-hour meter An instrument that monitors current with time. The indication is the product of current (in amperes) and time (in hours).

Amorphous Pertaining to a solid that is non-crystalline, having neither definite form nor structure, for example, amorphous silicon.

Antireflection coating A thin coating of a material applied to a solar cell surface that reduces the light reflection and increases light transmission.

Array current The electrical current produced by a photovoltaic array when it is exposed to sunlight.

Array operating voltage The voltage produced by a photovoltaic array when exposed to sunlight and connected to a load.

Balance of system Represents all components and costs other than the photovoltaic modules/array. It includes design costs, land, site preparation, system installation, support structures, power conditioning, operation and maintenance costs, indirect storage, and related costs.

Battery A storage device in which the electricity generated from the solar panel is stored.

Battery cycle life The number of cycles, to a specified depth of discharge, that a cell or battery can undergo before failing to meet its specified capacity or efficiency performance criteria.

Blocking diode An electronic component, which prevents the discharge of batteries through a PV panel.

Building-integrated photovoltaics (BIPV) The design and integration of PV technology into the building envelope, typically replacing conventional building materials.

Cathodic protection A method of preventing oxidation of the exposed metal in structures by imposing a small electrical voltage between the structure and the ground.

Charge controller An electronic component that controls the rate of charging of the battery and protects the battery from overcharging or deep-discharging.

Clean development mechanism (CDM) A mechanism of the Kyoto Protocol for reducing emissions through implementing projects in developing countries.

Concentrator A photovoltaic module, which includes optical components such as lenses (Fresnel lens) to direct and concentrate sunlight onto a solar cell of smaller area.

Conversion efficiency (solar cell) The ratio of electrical energy produced by a solar cell (under full sun conditions) to the energy from sunlight incident upon the solar cell.

Copper indium diselenide (CIS) A polycrystalline thin-film photovoltaic material (sometimes incorporating gallium (CIGS) and/or sulphur).

Crystalline silicon A type of photovoltaic cell made from a slice of single-crystal silicon or polycrystalline silicon.

Czochralski process (CZ) A method of growing large size, high quality semiconductor crystal by slowly lifting a seed crystal from a molten bath of the material under careful cooling conditions.

Deep cycle battery A battery with large plates that can withstand many discharges to a low state of charge.

Deep discharge Discharging a battery to 20% or less of its full charge capacity.

Depth of discharge (DOD) The ampere-hour removed from a fully charged cell or battery, expressed as a percentage of rated capacity. For example, the removal of 25 ampere-hours from a fully charged 100 ampere-hour rated cell results in a 25% depth of discharge.

Design month For the purpose of sizing an SPV system, it is necessary to choose the 'worst month(s)' (during which the solar radiation is not at its full; monsoon months or winter months) for which the system must meet the load requirements. This is termed as the design month.

Diffuse insolation Sunlight received indirectly as a result of scattering due to clouds, fog, haze, dust, or other obstructions in the atmosphere. Opposite of direct insolation.

Diffuse radiation Radiation received from the sun after reflection and scattering by the atmosphere and ground.

Direct current (DC) A type of electricity transmission and distribution by which electricity flows in one direction through the conductor, usually relatively low voltage and high current.

Direct radiation Solar radiation transmitted directly through the atmosphere.

Discharge rate The rate, usually expressed in amperes or time, at which electrical current is taken from the battery.

Feed-in-tariff The price per unit of electricity that a utility or a supplier has to pay for renewable electricity from private generators. The government generally regulates the tariff rate.

Flat-plate photovoltaic (PV) A PV array or module that consists of non-concentrating elements. Flat-plate arrays and modules use direct and diffuse sunlight.

Float charge The voltage required to counteract the self-discharge of the battery at a certain temperature.

Float-zone process (FZ) A method of growing a large-size, high-quality crystal, whereby coils heat a polycrystalline ingot placed atop a single-crystal seed. As the coils are slowly raised the molten interface beneath the coils becomes a single crystal.

Full sun The amount of power density in sunlight received at the earth's surface at noon on a clear day (about 1000 W/m^2).

Global irradiance The sum of diffuse and direct solar irradiance–the total irradiation (sunlight intensity) falling on a surface.

Global warming An overall increase in world temperature that may be caused by additional heat being trapped by greenhouse gases.

Grid-connected system A solar electric or PV system in which the PV array acts like a central generating plant, supplying power to the grid.

Grid-interactive system Same as grid-connected system.

Hybrid system A solar electric or photovoltaic system that includes other sources of electricity generation, such as wind or diesel generators.

Insolation The solar power density incident on a surface of stated area and orientation, usually expressed as watts per square metre or Btu per square foot per hour.

Inverters Electronic components that convert the DC to AC, to operate AC loads, or vice-versa.

Internal rate of return Measures the attractiveness of an investment decision.

Irradiance The direct, diffuse, and reflected solar radiation that strikes a surface. Usually expressed in kilowatts per square metre. Irradiance multiplied by time equals insolation.

Junction box A PV generator junction box is an enclosure on the module where PV strings are electrically connected and where protection devices can be located, if necessary.

Kilowatt (kW) A standard unit of electrical power equal to 1000 watts, or to the energy consumption at a rate of 1000 joules per second.

Kilowatt-hour (kWh) Thousand watts acting over a period of 1 hour. The kWh is a unit of energy, 1 kWh=3600 kJ.

Kilowatt peak (kW$_p$) Power output of a PV module under standard test conditions.

Lead–acid battery A general category that includes batteries with plates made of pure lead, lead–antimony, or lead–calcium immersed in an acid electrolyte.

Life cycle cost The estimated cost of owning and operating a PV system for the period of its useful life.

Load Any device or appliance that uses electrical power (solar electricity in case of a PV system)

Maximum power point tracker (MPPT) Impedance matching electronics used to hold the output of the PV array at its maximum value. Used with PV pumps, telecommunications, railway signalling, and so on.

Maintenance-free battery A sealed battery to which water cannot be added to maintain electrolyte level.

Megawatt (MW) Thousand kilowatts, or 1 million watts; standard measure of electric power.

Megawatt-hour Thousand kilowatt-hours, or 1 million watt-hours.

Modified sine wave A waveform that has at least three states (positive, off, and negative).

Mounting structure Metallic structures/frames used for mounting of PV components.

Net metering It allows power users to remain connected to the grid, and buy and sell power.

Nominal voltage A reference voltage used to describe batteries, modules, or systems (a 12-volt or 24-volt battery, module, or system).

Normal operating cell temperature (NOCT) The estimated temperature of a photovoltaic module when operating under 800 W/m^2 irradiance, $20\,°\text{C}$ ambient temperature and wind speed of 1 metre per second. NOCT is used to estimate the nominal operating temperature of a module in its working environment.

N-type Negative semiconductor material with more electrons than holes; current is carried through it by the flow of electrons.

NPV The difference between the PVC (present value of costs) and the PVB (present value of benefits).

Open circuit voltage The maximum possible voltage across a photovoltaic cell; the voltage across the cell in sunlight when no current is flowing.

Parallel connection A method of interconnecting two or more electricity producing devices, or power using devices, such that the voltage produced or required is not increased, but the current is additive. It is opposite of series connection.

Peak sun hours The equivalent number of hours per day when solar irradiance averages 1000 W/m². For example, six peak sun hours means that the energy received during total daylight hours equals the energy that would have been received had the irradiance for six hours been 1000 W/m².

Peak watt A unit used to rate the performance of solar cells, modules, or arrays; the maximum nominal output of a photovoltaic device, in watts (W_P) under standardized test conditions, usually 1000 watts per square metre of sunlight with other conditions, such as temperature specified.

Photovoltaic effect The phenomenon that occurs when photons, the 'particles' in a beam of light, knock electrons loose from the atoms they strike. When this property of light is combined with the properties of semiconductors, electrons flow in one direction across a junction, setting up a voltage. With the addition of circuitry, current will flow and electric power will be available.

Photovoltaic system A complete set of components for converting sunlight into electricity by the photoelectric process, including solar array and balance of systems.

Polycrystalline silicon (cell) Silicon solidified at such a rate that many small crystals have formed. The atoms within a single crystal are symmetrically arrayed, whereas in polysilicon crystals they are jumbled together.

Power conditioning equipment Electrical equipment, or power electronics, used in converting power from a photovoltaic array into a form suitable for subsequent use. A collective term for inverter, converter, battery charge regulator, and blocking diode.

PV array An interconnected system of PV modules that functions as a single electricity-producing unit. The modules are assembled as a discrete structure with common support or mounting.

PV cell A semiconductor device that converts sunlight directly into electricity, also called solar cell. All PV cells produce DC.

PV module A number of PV cells electrically interconnected and mounted together, usually in a common sealed unit or panel of convenient size for shipping, handling, and assembling into arrays.

Pyranometer An instrument used for measuring global solar irradiance.

Repeater stations It is an automated radio station that extends the range of communication.

Ribbon PV cells A type of PV device made in a continuous process of pulling material from a molten bath of photovoltaic material, such as silicon, to form a thin sheet of material.

Sealed battery A battery with a captive electrolyte and a resealing vent cap, also called a valve-regulated battery. Electrolyte cannot be added.

Series connection A method of interconnecting devices that generate or use electricity so that the voltage, but not the current, is additive one to the other. It is the opposite of parallel connection.

Short circuit current The current flowing freely through an external circuit that has no load or resistance; the maximum current possible.

Silicon (Si) A semi-metallic chemical element that makes an excellent semiconductor material for photovoltaic devices. It is commonly found in sand and quartz (as the oxide).

Sine wave A waveform corresponding to a single-frequency periodic oscillation that can be mathematically represented as a function of amplitude versus angle in which the value of the curve at any point is equal to the sine of that angle.

Solar noon The time of the day, at a specific location, when the sun reaches its highest, apparent point in the sky; equal to true or due, geographic south.

Solar irradiation or insolation The energy received per unit area from the sun in a specified time period. [In this document, the time period is generally taken to be a day and the solar irradiation is expressed in MJ/m^2 per day or kWh/m^2 per day].

Stand-alone PV system An isolated PV system not connected to a grid; may or may not have storage. But most stand-alone applications require a battery or other form of storage.

Standard test conditions Conditions under which a module is typically tested in a laboratory.

Substrate The physical material upon which a photovoltaic cell is applied.

Temperature compensation A circuit that adjusts the charge controller activation points depending on battery temperature. This feature is recommended if the battery temperature is expected to vary more than ±5 °C from ambient temperature.

Thin film photovoltaic module A photovoltaic module constructed with sequential layers of thin film semiconductor materials.

Tilt angle The angle at which a photovoltaic array is set to face the sun relative to a horizontal position. The tilt angle can be set or adjusted to maximize seasonal or annual energy collection.

Tilt factor The ratio of solar irradiation incident on a tilted PV array to global irradiation. An angle at which the SPV panel is installed to get maximum solar irradiation.

Venture capital It is a funding invested or available for investment in an enterprise that offers the probability of profit along with the possibility of loss; previously known as risk capital.

Wafer A thin sheet of semiconductor (photovoltaic material) made by cutting it from a single crystal or ingot.

Watt The rate of energy transfer equivalent to one ampere under an electrical pressure of one volt. One watt equals 1/746 horsepower, or one joule per second. It is the product of voltage and current (amperage).

Watt-hour One watt-hour of electricity is consumed when one watt of power is used for a period of one hour.

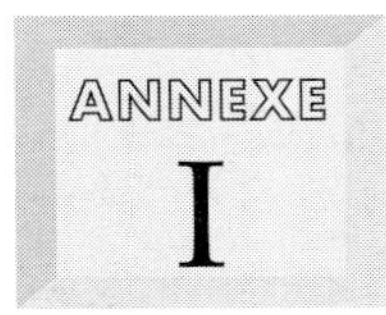

Solar PV manufacturers in India

Ammini Solar Pvt. Ltd
Plot No. 33-37
KINFRA Small Industries Park
St. Xaviers College PO
Trivandrum – 695 582
Phone: 0471-2705588
Fax: 0471-2705599
E-mail: solar@ ammini.com
Website: www.ammini.com

Bharat Electronics Ltd
116/2, Race Course Road
Jalahalli, Bangalore – 560 015
Phone : 080-25039300
Fax : 080-25039305
E-mail: imd@bel-india.com
Website: www.bel-india.com

Bharat Heavy Electricals Ltd
Electronics Division
Post Box No. 2606, Mysore Road
Bangalore – 560 026
Phone 080-26998553
Fax: 080-26744904, 26740137
E-mail: scpvbheledn.com
Website: www.bheledn.com

Central Electronics Ltd
4, Industrial Area
Sahibabad – 201 010
Phone : 0120-2895151
Fax: 01202-895148/42
E-mail: cel@celsolar.com
Website: www.celsolar.com

EMMVEE Solar Systems Pvt Ltd
55, 6th Main 11th Cross
Lakshmaiah Block, Ganganagar
Bangalore – 560 024

Phone: 080-23337427/28
Fax : 080-23332060
E-mail: emmvee@vsnl.com
Website: www.emmveesolar.com

Kotak Urja Pvt. Ltd
378, 10th Cross, 4th Phase, PIA
Bangalore – 560 058
Phone : 080-28363330
Fax : 080-28362347
E-mail: kotakurja@ gmail.com
Website: www.kotakurja.com

Maharishi Solar Technology Pvt. Ltd
A-14, Mohan Cooperative Industrial
Estate, Mathura Road
New Delhi – 110 044
Phone : 011-26959800 /26959701
Fax: 011-26959 669
E-mail: solar@maharishi.net
Website: www.maharishi.net

Moser Baer Photovoltaic Ltd
43B, Okhla Industrial Estate
New Delhi – 110 020
Phone 011-41635201-07, 26911570 –74
Fax : 011-41635211, 26911860
E-mail: pvinfo@moserbaer.in
Website: www.moserbaerpv.in

Photon Energy Systems
Plot No. 775 –K, Road No 45
Jublee Hills
Hyderabad – 500 033
Phone : 040-55661337/1338/1339
Fax: 040-55661340
E-mail: contact@photonsolar.com
Website: www.photonsolar.com

Premier Solar Systems (P) Ltd
41 & 42, Sri Venkateswara Cooperative
    Indl. Estate
Balanagar
Hyderabad – 500 037
Fax: 040-2271879
E-mail: premiersolar@yahoo.com

Rajasthan Electronics & Instruments Ltd
2, Kanakpura Industrial Area
Sirsi Road, Jaipur – 302 012
Phone : 0141-2203038
Fax: 0141-2202701/2352841
E-mail: reiljp@reiljp.com
Website: www.reiljp.com

Tata BP Solar India Ltd
Plot No. 78
Electronic City, Hosur Road
Bangalore – 560 100
Phone: 080-22358465
Fax: 080-28520972/28520116
E-mail: tatabp@ tatabp.com
Website: www.tatabpsolar.com

Titan Energy Systems Ltd
Aruna Enclave,Trimulgherry
Secunderabad – 500 015
Andhra Pradesh
Phone: 040-27791085/040-27790751
Fax: 040-27795629
E-mail: titan@ titansolar.com
Website: www.titansolar.com

USL Photovoltaics PVT Ltd
1/473 Avinashi Road, Neelambur
Coimbatore – 641 014
Phone : 0422-2627851
Fax : 0422-2628504
E-mail: udhaya@ vsnl.com
Website: www.solarpv.info

Udhaya Energy Photovoltaics (P) Ltd
1/279Z, Mudalipalayam
Arasur Post, Coimbatore – 641 407
Phone : 0422-2361170
Fax : 0422-2361180
E-mail: info@upvsolar.com
Website: www.upvsolar.com

Webel SL Energy Systems Ltd.
Plot No. N1, Block GP, Sector V
Salt Lake Electronic Complex
Kolkata – 700 091
Phone : 033-23578840
Fax: 033-23573258
E-mail: websol@ webelsolar.com
Website: www.webelsolar.com

XL Telecom Ltd
335 Chandralok Complex, SD Road
Secunderabad – 500 003
Phone : 040-27849094
Fax : 040-27840081
E-mail: info@ xltelecom.net
Website: www.xltelecom.net

Upcoming PV manufacturers
- Reliance Industries Limited
- MSPL
- Signet Solar India Limited
- Sharp Solar Systems
- Khandelwal Solar
- Solar Cube
- Prism Solar

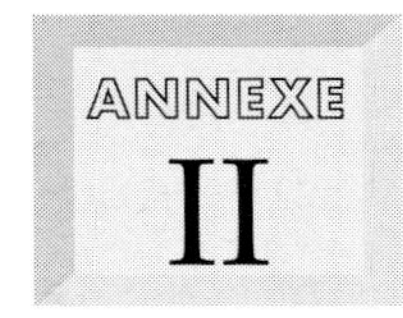

List of programme implementation agencies

ANDHRA PRADESH
Vice-chairman and Managing Director
**Non-Conventional Energy
  Development Corporation of Andhra
  Pradesh (NEDCAP) Ltd**
5-8-207/2, Pisgah Complex
Nampally, Hyderabad – 500 001
Tel: 040-23201172
Fax: 040-23201666
E-mail: nedcap@ap.nic.in
Website: http://www.nedcap.org/

ARUNACHAL PRADESH
Director
**Arunachal Pradesh Energy
  Development Agency**
Urja Bhawan Tadar Tang Marg
Post Box No. 141, Itanagar – 791 111
Tel: 0360-211160/216937
Fax: 0360-214426
Website: http://www.apedagency.com

ASSAM
Director
**Assam Energy Dev. Agency (under
  Science and Technology Deptt. Govt.
  of Assam)**
Bigyan Bhawan, Near IDBI Building
G S Road, Guwahati – 781 005
Tel: 0361-2464619, 2464621
Fax: 0361-2464617
E-mail: aeda@india.com
Website: http://www.assamrenewable.
  org

BIHAR
Director
**Bihar Renewable Energy Development
  Agency**

1st Floor, Sone Bhawan
Virchand Patel Marg
Patna – 800 001
Tel: 0612-2233572, 2220493 (R)
Fax: 0612-2228734
E-mail: dir_breda@sancharnet.in

CHHATTISGARH
Principal Secretary
Energy and Chief Executive Officer
**Chhattisgarh State Renewable
  Energy Development Agency
  (CREDA) Mantralaya**
DKS Bhawan, Raipur
Tel: 0771-2221308, 5080308
Fax: 0771-2221163
E-mail; vivekdhand@cg.nic.in
Website: http://www.credacg.com

Director
**Chhattisgarh State Renewable Energy
  Development Agency**
MIG/A-20/1, Sector 1
Shankar Nagar, Raipur
Tel: 0771-2426446; 5022050 (R)
Fax: 5066770
Email: creda@epatra.com
  credacg@rediffmail.com
  credacg@yahoomail.com
Website: http://www.credacg.com

DELHI
**Department of Energy Efficiency &
  Renewable Energy Management
  Centre**
SLDC Building, Minto Road
New Delhi – 110 001
Tel: 011-23238263/23234994

GOA
Member Secretary
Goa Energy Development Agency
DST&E Building, 1st Floor
Saligo Plateau, Opp. Seminary
Saligao, Bardez, Goa – 403 511
Tel: 0832-271194

GUJARAT
Director
**Gujarat Energy Development Agency
  (GEDA)**
4th Floor, Block No.11 & 12
Udyog Bhawan, Sector 11
Gandhi Nagar – 382 017
Gujarat
Tel: 079-23247086/89
Fax: 079-23247090
E.mail: info@geda.org.in
Website: http://www.geda.org.in

HARYANA
Director
**Haryana Renewal Energy Development
  Agency (HAREDA)**
SCO 48, Sector 26
Chandigarh – 160 019
Tel: 0172- 2791917, 2790918,
  2790911(O), 2794185 (R)
Fax: 0172-2790928
Email : hareda@chd.nic.in
Website: http://www.hareda.gov.in

HIMACHAL PRADESH
Chief Executive
HIMURJA
SDA Complex, Kasumpti
Shimla – 171 009
Tel: 0177-2620365 (O) 2620371 (R)
Fax: 0177-2620365
E Mail: himurja@hp.nic.in
Website: http://himurja.nic.in

JAMMU AND KASHMIR
Chief Executive Officer
**Jammu & Kashmir Energy
  Development Agency (JAKEDA)**
12 BC Road, Rehari, Jammu
Tel: 0191-2546495, 2552725(R)
Fax: 0191-2546495

JHARKHAND
Director
**Jharkhand Renewable Energy
  Development Agency**
328 B, Road No.4, Ashok Nagar
Ranchi – 834 002
Tel: 0651-2246970
Fax: 0651-2240665
Email : info@jreda.com
Website: http://www.jreda.com

KARNATAKA
Managing Director
**Karnataka Renewable Energy
  Development Agency Ltd**
19, Maj. Gen. A D Loganadan
INA Cross, Queen's Road
Bangalore – 560 052
Tel: 080- 282220 (O), 23365590 (R)
Fax: 080-22257399
Email: drbsmdkredl@yahoo.com
Website: http://kredl.kar.nic.in

KERALA
Director
**Agency for Non-Conventional Energy
  and Rural Technology (ANERT)**
PATTOM POPB No.1094
Kesavadasapuram
Thiruvananthapuram – 695 004
Tel: 0471- 2440121, 2440122, 2440124
Fax: 0471-2449854
Email: anert@vsnl.com
Website: http://education.vsnl.com/
  anert/

MADHYA PRADESH
Managing Director
MP Urja Vikas Nigam Ltd Urja Bhawan
Main Road No. 2, Shivaji Nagar
Bhopal – 462 016
Tel: 0755-2556245, 2553595
Fax: 0755-2556245
Email: mpuvn@sancharnet.in
Website: http://www.mprenewable.org

MAHARASHTRA
Director General
**Maharashtra Energy Development
  Agency (MEDA)**
S. No. 191/A, Phase1, 2nd Floor

MHADA Commercial Complex
Opp. Tridal Nagar, Yerawada
Pune – 411 006
Tel: 020-26614393, 26614403, 26615322
Fax: 020-26615032
Email : meda@vsnl.com
  mahaurjamumbai@yahoo.com
  dg.meda@nic.in
Website: http://www.mahaurja.com

MANIPUR
Director
**Manipur Renewable Energy
  Development Agency (MANIREDA)**
Department of Science, Technology Sai
  Road, Takyelpat
Imphal – 795 001
Tel: 0385-2222685
Fax: 0385-224930
Email: mlou_singh@yahoo.com

MEGHALAYA
Director
**Meghalaya Non-conventional & Rural
  Energy Development Agency**
Lower Lachaumiere, Opp. P&T
  Dispensary, Near BSF Camp (Mawpat)
Shillong – 793 001
Telefax: 0364-537343
Website: http://mnreda.gov.in

MIZORAM
Director
**Zoram Energy Development Agency
  (ZEDA)**
Zuangtui P.O. Zemabawk
Aizawl – 796 017
Tel: 0389-2350664, 2350664, 2350665
Fax: 0389-2350664/2350664
Email: zedaizawi@hotmail.com

NAGALAND
Project Director
**Nagaland Renewable Energy
  Development Agency (NREDA)**
NRSE Cell Rural Development
  Department
Nagaland Secretariat, Kohima
Telefax: 0370-241408

ORISSA
Chief Executive Officer
**Orissa Renewable Energy Development
  Agency**
S-59, Mancheswar Industrial Estate
Bhubaneswar – 751 010
Tel: 0674-2580660
Fax: 0674-2586368
Website: http://www.oredaorissa.com

PUNJAB
Chief Executive
Punjab Energy Development Agency
SCO 134-136, Sector 34-A
Chandigarh – 160 036
Tel: 0172-2663392; EPBX 2663328/3382
Fax: 0172-2646384
Email: peda@glide.net.in
  mpsingh66@yahoo.com
  peda_spa@yahoo.co.in

Director
Punjab Energy Development Agency
SCO 134-136, Sector 34-A
Chandigarh – 160 036
Tel: 0172-663392; EPBX 663328/663382
Fax: 0172-2646384
Email : peda@glide.net.in

RAJASTHAN
Chairman & Managing Director
**Rajasthan Renewable Energy
  Corporation Limited**
E-166, Yudhister Marg
'C' Scheme, Jaipur – 302 001
Tel: 0141-2225898/2228198 (O)
Fax: 0141-2226028
Email : rrec_jai@yahoo.co.in
  gsomanl@datainfosys.net
Website: http://www.rajenergy.com

SIKKIM
Director
**Sikkim Renewable Energy
  Development Agency**
Government of Sikkim
D.P.H. Road (Near Janta Bhawan)
Gangtok – 737 101
Tel: 03592- 22659
Fax: 03592-22245
Email: slg_sreda@sancharnet.in

TAMIL NADU
Chairman & Managing Director
**Tamil Nadu Energy Development
 Agency (TEDA)**
EVK Sampath Building
Chennai 600 006
Tel: 044-28224832
Fax: 044-28236592
Email: teda@md4.vsnl.net.in

TRIPURA
Chief Executive Officer
**Tripura Renewable Energy
 Development Agency**
Vigyan Bhawan, 2nd Floor
Pandit Nehru Complex
West Tripura, Agartala – 799 006
Tel: 0381-2225900, 2326139
Fax. 0381-225900
Email: tredaagt@yahoo.co.in

UTTAR PRADESH
Chairman and Director
**Non-conventional Energy
 Development Agency (NEDA)**
Vibhuti Khand, Gomti Nagar
Lucknow – 226 010
Tel: 0522-2720652
Fax: 0522-2720779, 2720829
Email : nedaup@dataone.in
Website: http://neda.up.nic.in
 http://www.uttara.in/ureda/intro.
 html

WEST BENGAL
Director
**West Bengal Renewable Energy
 Development Agency**
Bikalpa Shakti Bhawan
Plot- J-1/10, EP & GP Block
Salt Lake Electronics Complex
Sector V, Kolkata – 700 091
Tel: 033-23575038, 23575348
Fax: 033-23575037, 23575347
Email: wbreda@cal.vsnl.net.in
Website: http://www.wbreda.org

ANDAMAN & NICOBAR ISLANDS
Executive Engineer
(NRSE) Electricity Department
Office of the Executive Engineer NRSE

Division, Prothrapur, Port Blair – 744 105
Tel: 03192-250577
Fax: 03192-250930

CHANDIGARH
Director (Science & Technology)
Chandigarh Administration
Additional Town Hall Building
2nd Floor, Sector-17 C, Chandigarh
Tel: 0172-2745502 (O)
Fax: 0172-2726698

DADRA & NAGAR HAVELI
Development and Planning Officer
Administration of Dadra & Nagar Haveli
Silvassa
Tel. 0260-642070

LAKSHADWEEP
Executive Engineer
Electricity Department
Lakshadweep Administration
Kavaratti – 682555
Tel: 04896-262363
Fax: 04896-262936

PUDUCHERRY
Project Director (IREP)
**Renewable Energy Agency of
 Puducherry**
No.10, Second Main Road
Elango Nagar, Puducherry
Tel: 0413-2244219 (O)
Fax: 0413-333601

LADAKH
Chairman & Chief Executive Councillor
**Ladakh Autonomous Hill Development
 Council (LAHDC)**
LEH, Ladakh – 194 101
J & K State

IREDA
**Indian Renewable Energy
 Development Agency Ltd.**
Core - 4 -A, East Court, First Floor,
IHC Complex
Lodhi Road, New Delhi – 110 003
Fax 91-11-24602855
Email: gen@ired.1.globemail.com
www.iredaltd.com

RE-related international organizations in India

DFID (Department for International Development)

BHC (British High Commission)

ECN (Energy Research Centre of the Netherlands)

European Commission

Ford Foundation

Global Development Network

GTZ (German Technical Cooperation)

IDRC (International Development Research Centre)

International Global Energy Strategies

Intermediate Technology Development Group

IT Power India Ltd

NEDO (New Energy and Industrial Technology Development Organization)

Overseas Development Institute

SIDA (Swedish International Development Agency)

SDC (Swiss Agency for Development and Cooperation)

US EPA (Environmental Protection Agency)

The World Bank

UN Department of Economic and Social Affairs

United Nations Development Programme

United Nations International Development Organization

World Health Organization

IV

List of important solar PV-related websites

Affiliation	Website
Ministry of New and Renewable Energy	www.mnre.gov.in
Indian Renewable Energy Development Agency	www.ireda.in
Indian Solar PV Industry	Pl. refer to websites mentioned in Annexe I
TERI (The Energy and Resources Institute)	www.teri.res.in
Renewable Energy World	www.renewableenergyworld.com
Solar Plaza	www.solarplaza.com
Solar Buzz	www.solarbuzz.com
PV Tech	www.pv-tech.org
American Solar Energy Society	www.ases.org
Eurosolar	www.eurosolar.de
Renewable Energy Network	www.ren21.net
Solar Electric Light Fund	www.self.org
Sandia National Laboratories	www.sandia.gov
National Renewable Energy Laboratory	www.nrel.gov
European Photovoltaic Industry Association	www.epia.org
International Solar Energy Society	www.ises.org
Solar Electric Power Association	www.solarelectricpower.org
Find Solar	www.findsolar.com

Index